The Science of Green Energy

Concern for the environment and the impacts of pollution have brought about the need to shift from the use and reliance on hydrocarbons to energy-power sources that are pollution neutral or near pollution neutral or renewable. Moreover, the impact of 200 years of industrialization and surging population growth threatens to exceed the future supply of hydrocarbon power sources. Therefore, the implementation of green energy sources is surging. *The Science of Green Energy* presents technologies and techniques, as well as real-world usage of and operation of today's green energy-based applications. This practical book is designed to be used as an information source for the general reader, or for a course in energy, chemistry, or in renewable energy engineering fields where green energy is becoming a key player. It is intended to fill a wide gap of missing information in published texts dealing with the green energy revolution currently in progress; it specifically provides information involving the many different sources of energy.

- Provides an understanding of the current science buttressing green energy sources and the possible health and safety issues associated with each of the green energy types.
- Explains the complexity and limitations of green energy in basic and understandable terms for almost anyone and everyone.

Frank R. Spellman, PhD, is a retired assistant professor of environmental health at Old Dominion University, Norfolk, Virginia, and the author of more than 160 books covering topics ranging from concentrated animal feeding operations (CAFOs) to all areas of environmental science and occupational health. He consults on homeland security vulnerability assessments for critical infrastructures, including water/wastewater facilities, and conducts audits for Occupational Safety and Health Administration and the Environmental Protection Agency inspections throughout the country. Dr. Spellman lectures on sewage treatment, water treatment, and homeland security, as well as on safety topics, throughout the country and teaches water/wastewater operator short courses at Virginia Tech in Blacksburg, Virginia.

The Science of Green Energy

Frank R. Spellman

CRC Press
Taylor & Francis Group
Boca Raton London New York

CRC Press is an imprint of the
Taylor & Francis Group, an **informa** business

Cover credit: Shutterstock

First edition published 2024
by CRC Press
6000 Broken Sound Parkway NW, Suite 300, Boca Raton, FL 33487-2742

and by CRC Press
4 Park Square, Milton Park, Abingdon, Oxon, OX14 4RN

CRC Press is an imprint of Taylor & Francis Group, LLC

© 2024 Taylor & Francis Group, LLC

ISBN: 978-1-032-57365-6
ISBN: 978-1-032-57366-3
ISBN: 978-1-003-43905-9

DOI: 10.1201/9781003439059

Typeset in Times
by Newgen Publishing UK

Contents

Preface

The Science of Green Energy is the twelfth volume in the acclaimed series that includes *The Science of Carbon Capture and Storage* (in production), *The Science of Lithium (Li)*, *The Science of Ocean Pollution*, *The Science of Electric Vehicles (EVs): Concepts and Applications*, *The Science of Rare Earth Elements: Concepts and Applications*, *The Science of Water*, *The Science of Air*, *The Science of Environmental Pollution*, *The Science of Renewable Energy*, *The Science of Waste*, and *The Science of Wind Power*, all of which bring this highly successful series fully into the twenty-first century. *The Science of Green Energy* continues the series mantra based on good science and not feel-good science. It also continues to be presented in the author's trademark conversational style—making sure communication is certain, not a failure.

This practical, direct book presents technologies and techniques, as well as real-world usage of and operation of today's green energy-based applications. This book is designed to be used as an information source for the general reader, or for a course in energy, chemistry, or in renewable energy engineering fields where green energy is becoming a key player.

Concern for the environment and the impact of environmental pollution has brought about the trend (and the need) to shift from the use and reliance on hydrocarbons to energy-power sources that are pollution neutral or near pollution neutral and renewable. We are beginning to realize that we are responsible for much of the environmental degradation of the past and present—all of which is readily apparent today—human footprints and other environmental footprints grow with each passing day. Moreover, the impact of more than 200 years of industrialization and rising population growth has far exceeded the future supply of hydrocarbon power sources. So, the implementation of green energy sources is surging.

This book is about a serious, growing problem related to unmanaged commons. The unmanaged commons addressed herein is the unmanaged polluting of our common environment, the unmanaged commons. In *The Science of Green Energy*, the facts about the unmanaged commons being polluted every day, every hour, every minute, and every second are presented in my characteristic no-holds-barred, no limits, no political leanings, no nutcase wacko, off-the-wall bologna: Just the facts, ma'am—and everyone else (Again, thanks, Sgt. Joe Friday).

You may have heard one of these old sayings, idioms, or phrases, that green energy (excluding the initial capital outlay and infrastructure expense) is one of the best things in Nature that is free; think green and live green; energy as spotless as a breeze; being green is remaining clean; let's add green energy to life; and finally, live green, love green, think green! Having spent most of my life studying energy, analyzing energy, writing about energy, and teaching environmental health and engineering students about energy, I have come to realize, early on, that energy is indeed the shot in the arm, so to speak, for maintaining the so-called good life on Earth.

The irony in this statement about green energy being the shot in the arm, the tonic for maintaining life on Earth, is apparent when explained to humans who normally do not think about green energy. Many want energy to fuel their lives, but also want energy that does not contaminate the "unmanaged" environment. The opponents of Green Energy feel that green energy-produced electricity is not the answer. This is like the mindset of many who have a negative view, for example, of electric vehicles because of range anxiety, lack of charging availability, and the cost of electric vehicles. Moreover, since today's electric vehicle batteries are not of the change-out variety, their life span over the years of driving and recharging is limited. Thus, when the battery is exhausted and can't be recharged, what is to be done with the vehicle? In short, to make electric vehicles powered by green energy storage units (batteries) viable, they must be powered by batteries that can be changed out with a freshly charged battery. This would allow the driver to motor on simply by pulling into a gas station and changing out the near-dead battery with one that is freshly charged.

But keep in mind that the energy needed to charge an electric vehicle and other electricity-powered devices comes from either hydrocarbon fuels or from renewable energy sources (aka Green Energy).

Yes. The average person learns, usually early on, that energy is a must, a must-have, a crucial factor for maintaining his or her good health, lifestyle, and life.

The bottom line: even though human welfare is intricately linked, interconnected with the usage of hydrocarbon natural resources, humans have substantially altered the face of the environment within only a few centuries. The face of today and tomorrow's environment is quite apparent, obvious, visible; it's Green—or it should be.

One area that is often ignored when discussing green energy is green energy jobs and workers. Green energy is viewed by many as a good thing, but it is less under-stood that operation of green energy systems also includes potential exposure to on-the-job health and safety hazards. In this book, green energy jobs are described, including making living and working spaces powered by green energy and their potential hazards.

F.R. Spellman, Norfolk, VA

Prologue[1]

Eventually, growth in the globe's population and material economy will confront humanity. When this occurs (and it will), we will either adjust and survive or we will simply join the ranks of the dinosaurs, Dodo birds, Passenger Pigeons, Golden Toads, or those other several species presently experiencing the Sixth Extinction.

–Frank R. Spellman, 2012

During a recent Rabbit and Grasshopper conversation:

Grasshopper stated to his friend Rabbit: "To fix human energy, population, unemployment and many of their economic problems, what they need right now, my friend, is innovation, innovation, innovation …"

After deliberate and well-practiced thumping of his foot, Rabbit replied to his friend Grasshopper: "No, my long-legged friend, … to fix all of humankind's economic problems, what they need first is discovery, discovery, discovery and then invention, invention, invention … followed up by innovation, innovation, innovation … Hmmmm, human leadership, brain power, common sense, and accountability would also help."

Grasshopper: hmmmmmmmmmmm, buzzzzzzzz, etc., "Well, they ain't too smart … them humans … All they need to do is ask us. We know how to economize … and how to do the rest."

Rabbit replied: "Right on, Grasshopper!"

–A friend of all grasshoppers and rabbits, 2014

NOTE

1 Much of the information in this section is from F.R. Spellman (2015) *Economics for the Environmental Professional*. Boca Raton, FL: CRC Press.

About the Author

 **Frank R. Spellman, CSP, CHMM, PhD**, is a retired US naval officer with 26 years active duty, and also a retired full-time adjunct assistant professor of environmental health at Old Dominion University, Norfolk, Virginia. He is the author of more than 161 books covering topics ranging from a 15-volume homeland security series, to several safety, industrial hygiene, stormwater management, air pollution, and security manuals, and also including concentrated animal feeding operations (CAFOs), to all areas of environmental science and occupational health and regulatory compliance. He has also authored 20 novels. Many of his texts are readily available online at Amazon.com and Barnes and Noble.com, and several have been adopted for classroom use at major universities throughout the United States, Canada, Europe, and Russia; two have been translated into Chinese, Japanese, Arabic, Korean, and Spanish for overseas markets. Dr. Spellman has been cited in more than 1,850 publications. He serves as a professional expert witness for three law groups and as an incident/accident investigator and security expert for the US Department of Justice and a northern Virginia law firm. In addition, he consults on homeland security vulnerability assessments for critical infrastructures including water/wastewater facilities nationwide and conducts pre-Occupational Safety and Health Administration (OSHA)/Environmental Protection Agency (EPA) audits throughout the country. Dr. Spellman receives frequent requests to coauthor with well-recognized experts in several scientific fields; for example, he is a contributing author of the prestigious text *The Engineering Handbook,* 2nd ed. (CRC Press). Dr. Spellman lectures on wastewater treatment, water treatment, and homeland security, and lectures on safety topics throughout the country and teaches water/wastewater operator/regulatory short courses at Virginia Tech (Blacksburg, Virginia). In 2011–2012, he traced and documented the ancient water distribution system at Machu Pichu, Peru, and surveyed several drinking water resources in Amazonia-Coco, Ecuador. Dr. Spellman also studied and surveyed two separate potable water supplies in the Galapagos Islands; he also studied and researched and studied Darwin's finches while in the Galapagos. He holds a BA in public administration, a BS in business management, an MBA, and an MS and PhD in environmental engineering.

Part I

The Basic

1 Going Green

INTRODUCTION

Why a text on the Science of Green Energy?

Good question.

Studying green energy through science affords us the opportunity to not only grasp the pressing need for green (aka renewable) energy, but also the complicated task of developing viable energy sources for the future.

Regarding future energy needs, there is no question or doubt that we are facing an international crisis of unprecedented proportion. We are headed for a train wreck: the end of affordable and accessible petroleum products, and the resulting economic and political turmoil. Keep in mind that we not only need to replace liquid hydrocarbon fuels because the wells are almost dry, but also that we need to clean up the environment—we need to produce reliable green/renewable energy that will not destroy or pollute our fragile environment. Even though peak oil is here, and hydrocarbon energy resources are soon to be strained to the limit, and the train wreck of world economies collapsing will then move beyond a pending disaster to actual disaster itself, we need to follow Adrienne Rich's (1973) perfect words and "dive into the wreck." Simply, a substantial and reliable supply of green energy—a subset of renewable energy—is the wreck-preventer we need. That is, the train wreck can be avoided simply by pursuing emerging technologies that include making oil from garbage, algae, plastics, sewage, and agricultural and forestry biomass/waste to power fuel cells and full-sized plug-in diesel hybrid SUVs. Moreover, the development of these emerging technologies must be accompanied by advancements in solar, wind, and ocean technologies.

SIDEBAR 1.1—PEAK OIL

According to the US Department of Energy (USDOE, 2007), the *peak oil theory* states that the world's oil production rate will reach a maximum rate and then enter terminal decline. This concept is based on the observed production rates of individual oil wells, and the combined production rate of a field or related oil wells. This is not to say that the world is running out of oil, but it does signal the end of abundant and cheap oil on which all industrial nations depend (Campbell & Laherrere, 1998). Peak

DOI: 10.1201/9781003439059-2

oil, also known as the Hubbert peak theory, concerns the long-term rate of conventional oil (and other fossil fuels) extraction and depletion. M.K. Hubbert proposed, in a 1956 paper he presented at a meeting of the American Petroleum Institute, that oil production in the United States would peak between 1965 and 1970 (Hubbert, 1956). US oil production peaked in 1971 and has been decreasing since then. Hubbert's theory is subject to continued discussion because of the potential effects of lowered oil production, and because of the ongoing debate over aspects of energy policy. Opinions on the effect of passing Hubbert's peak range from faith that the market economy will produce a solution, assuming major investments in alternatives will occur before a crisis, to predictions of doomsday scenarios of a global economy unable to meet its energy needs (Gwyn, 2004). This thesis postulates that the price of oil at first will escalate and then retreat, as other types of fuel and energy sources are developed and used (CERA, 2006). Pessimistic predictions of future oil production operate on the thesis that the peak has already occurred (Deffeyes, 2007), that oil production is on the peak, or that it will occur shortly (Koppelaar, 2006).

Although many experts were skeptical of his prediction, Hubbert was proven correct when US oil production peaked in 1971 (Brandt, 2007). Hubbert's prediction remained accurate even after the discovery of the Prudhoe Bay oil field (after 1971) with its great volume of oil still not enough to bring US oil production out of a long-term decline (USDOE, 2007).

Peak oil production and subsequent global production decline (based on optimistic assessments) is forecast to begin by about 2020 or later. Keep in mind that the exact date of world oil peaking is a best guest; the actual date is not known with certainty, complicating the decision-making process. The decision-making process refers to deciding what or which alternative sources of energy to pursue to prevent the economic and social impacts of oil peaking from being chaotic. A basic problem in predicting oil peaking is uncertain and politically biased oil reserves claims from many oil-producing countries. In some instances, the veracity of such claims may be questionable (Hirsch, 2005).

If we accept predictions about peaking oil, the possible effects and consequences of peaking oil cannot be drawn and displayed without using a very wide brush. Few would argue against the fact that the widespread use of fossil fuels has been one of the most important agents of economic growth and prosperity since the Industrial Revolution. Fossil fuels have allowed humans to participate in the consumption of energy at a greater rate than it is being replaced. Some believe that mankind is in for a momentous paradigm shift when oil production decreases. This shift will cause humans and our modern technological society to change drastically. The actual impact of peak oil will depend heavily on the rate of decline and the development and adoption of effective alternatives. If alternatives are not forthcoming, the products produced with oil (including fertilizers, detergents, adhesives, solvents, and most plastics) will become scarce and expensive. In 2005, the USDOE published a report (known as the *Hirsch Report*) with a stark warning:

> The peaking of world oil production presents the U.S. and the world with an
> unprecedented risk management problem. As peaking is approached, liquid
> fuel prices and price volatility will increase dramatically, and, without timely

mitigation, the economic, social, and political costs will be unprecedented. Viable mitigation options exist on both the supply and demand sides, but to have substantial impact, they must be initiated more than a decade in advance of peaking.

(USDOE, 2005, Intro.)

Many of us have come to realize that we pay a price for what we call "the good life." Consumption of the world's resources makes us all at least partially responsible for the shortages and the resulting environmental pollution. Simply, it is a matter of record, proven, that pollution (and its ramifications) is one of the inevitable products of "the good life" when maintained by fossil fuels. But fuel shortages and pollution emanating from fossil fuel usage is not something any single individual causes or can totally prevent or correct. To reduce fuel shortages and its harmful effects, we must band together as an informed, knowledgeable electorate and pressure our elected decision-makers to manage the problem now, for the future. We need to replace fossil fuel with clean and efficient renewable energy. Keep in mind, however, we need to do this slowly. Change takes time and we must ensure not only energy availability in the future, but also that we have enough energy to support our needs. We need to quiet that vocal minority: the doomsayers and naysayers who shout out that the transition from our current fossil fuel-based economy to an economy based entirely on renewable energy would be too "costly" and "impractical." The present text was written not to silence these doubting Thomases but instead to make their arguments hollow and, at best, moot. We have an infinite source of green energy but only a limited amount of fossil fuel.

Throughout this text, commonsense approaches and practical (and sometimes poetic) examples are presented. Because this is a science text, we have adhered to scientific principles, models, and observations. However, you need not be a scientist to understand the principles and concepts we present—we go easy on the hard math and science and present the material in a user-friendly manner. There is no failure to communicate. However, what you really need is an open mind, a love for the challenge of wading through the muck, an ability to decipher problems, and the patience to answer questions, maybe only for yourself, that are relevant to each topic presented. Real-life situations are woven throughout the fabric of this text, and presented in straightforward, plain English, to equip you with the facts, knowledge, and information you need in order to understand complex issues and to make your own informed decisions.

As a companion text to *The Science of ...* books in this series, this book follows the same proven format used in the other ten texts. But like its forerunners, this text is not an answer book. Instead, it is designed to stimulate thought. Although solutions to specific green energy questions are provided, we also point out the disadvantages and hurdles that need to be overcome to make renewable energy a viable alternative to fossil fuels. Our goal herein is to provide the framework of principles you can use to understand the complexity of substituting green energy sources for fossil fuels.

The Science of Green Energy and Practices is designed to reach a wide range of diverse reader and student backgrounds. The text focuses on green energy derived from natural processes that are replenished constantly in various forms emanating from sun, wind, water (tides and waves), geothermal, biomass, hydroelectricity, biofuels, and hydrogen fuel cells designed and implemented to prevent pollution of the atmosphere, of surface and groundwater, and of soil (the three environmental mediums). Keep in mind that all these naturally produced processes are critical to and intended to sustain us—to aid in our very survival. Because green energy and pollution prevention are real-world issues or problems, it logically follows that we can solve these issues and/or problems by using real-world methods—that's what *The Science of Green Energy* is all about.

WHAT IS GREEN ENERGY?

The move toward green energy is being driven by a relatively new awareness of what we are doing to our environment, that is, our impact on our surroundings. Moreover, because of this awareness of our impact on the environment, the importance of transition from conventional energy resources to renewable sources—green energy sources—has become clear to the average person, the consumer. The movement toward a sustainable environment is thought to be possible when we switch from conventional energy sources (namely, from hydrocarbon sources) to green energy sources (aka green power). The problem with the green energy movement is that many commonly use it to refer only to renewables, and others are confused by the term.

So, to clear up confusion and misunderstanding of the term "green energy," we can simply say that it most commonly refers to the US market, where it provides a range of services and products with green energy. Keep in mind that the term green energy is not to be confused with the term renewable "energy." Unfortunately, and incorrectly, the two terms are often confusing and used interchangeably.

Anyway, green energy is any energy type that is generated from natural resources, such as water, wind, or sunlight. The confusion between green energy and renewable energy is that green energy comes from renewable energy sources, but there are some differences between renewable and green energy.

Keep in mind that these two energy resources do not harm the environment, that is, regarding releasing greenhouse gas emissions into the atmosphere.

Again, as a source of energy, green energy often comes from renewable energy technologies such as solar energy, wind power, biomass, geothermal, and hydroelectric power. Note that these technologies work in different ways, whether taking power from the sun, as with solar panels, or using the flow of water, or using wind turbines to generate energy.

The key point here is that to be regarded as green energy, a resource can't produce pollution. Of course, this is not the case with fossil fuels. What this really means is that not all sources of renewable energy are green. A good example is power generation that burns organic material from sustainable forests. This may be renewable, but it is not necessarily green due to the carbon dioxide (CO_2) produced by the burning process itself.

With green energy sources, it is all about many of them being able to naturally replenish, whereas fossil fuel sources like natural gas or coal can't be replenished in quick order—millions of years are required to renew these sources. A huge advantage of green energy sources is that they do not disturb ecosystems by drilling or mining.

Now, let's shift gears to regard future energy needs. Keep in mind that if we are to shift totally to producing battery-powered vehicles versus internal-combustion engines, we need further development of the technology involved. Simply, we need to lessen range anxiety and the total end of operation of battery-powered vehicles because the batteries are integral and irreplaceable in the present configuration of battery-powered vehicles—that is, when the vehicle's battery dies, so does the vehicle—we need to develop batteries with greater range and accessible easy gas-station battery changeout capability to enable to continue operation of the vehicle whenever battery changeout is required.

These statements about EV range anxiety demonstrate how complicated, perplexing, confusing, befuddling, and confounding when/while we attempt to transition from hydrocarbon-fueled vehicles to electric-powered vehicles. Moreover, Umar Shakir (2023) points out that at the "present time EV owners are overall not satisfied with the reliability of the charging infrastructure available in the US [...] things are looking worse than last year."

HYPOTHETICAL CASE STUDY 1.1

Our select team of renewable energy experts is being given a $250 million dollar grant to do whatever it takes to develop a super battery or fuel cell to power the cars of the future. Accompanying specifications and directions simply state that this battery or fuel cell needs to be large capacity (large ampere-hour capacity) that can be continuously recharged via light striking super-sensitive solar sensors. To charge any rechargeable battery, a certain amount of current or electron flow is necessary. The solar supersensors we invented will produce substantial current flow. These super sensors are wired from the battery itself to at least four positions on the roof of our experimental automobile where they can constantly be exposed to daylight—any level of daylight. In addition, we know what you're thinking. You are thinking that it does get dark at night, so how are you going to continuously recharge the car's battery, fuel cell, or whatever else we choose to call it? Good question: it has a simple answer. Two additional wires and ultra-super-super light sensors (that we also invented) are to be run from the battery to the car's headlamps, one to each. Thus, during nighttime driving, the light from the headlamps will be sensed by the ultra-super-super sensors and sent back to the battery. Now, we know electrical engineers out there in la-la land are scratching their heads and wondering where did we get such an unworkable idea? We are not finished with our explanation ... hold onto your brain cells. An amplification or booster device (like a transistor that amplifies a signal but is many times more powerful) is placed within the wiring between the battery and the external and headlamp cell connections. We are not talking about perpetual motion here; we are talking about perpetual energy.

Is this scenario possible? Anything, absolutely anything is possible. We are Americans; we believe we can do anything if we put our minds to it and history has shown this to be true.

Now is the time.

Let's say that we can create a power plant arrangement consisting of a powerful battery or fuel cell with a built-in recharge system like the one just described. What's next? Another good question. What's next is the final ingredient in any innovative processes: education, training, and on-the-job training (OJT). To produce innovation and research and development grants, we must have personnel who are highly educated and trained. Unfortunately, in the United States at present, we are losing (or have already lost in some areas) our edge in higher education. The United States will not restore itself to preeminence in science, technology, engineering, and mathematics until we shift our focus, money, and talent to educate our youth in these critical subjects.

==

THE 411 ON ENERGY[1]

Defining energy can be accomplished by providing a technical definition or by a characterization in layman's terms. Because the purpose of this book is to reach technical readers as well as a wide range of general readers, definitions provided herein and hereafter are best described as technical–nontechnical-based. Consider the definition of energy, for example. It can be defined in several ways. In the broad sense, energy means the capacity of something, a person, an animal, or a physical system (machine), to do work and produce change. In layman's terms, energy is the amount of force or power that when applied, can move one object from one position to another. It can also be used to describe someone doing energetic things such as running, talking, and acting in a lively and vigorous way. It is used in science to describe how much potential a physical system must change. It also is used in economics to describe the part of the market where energy itself is harnessed and sold to consumers. For our purposes in this text, we simply define energy in technical–nontechnical-based terms as something that can do work or the capacity of a system to do work.

There are two basic forms of energy: kinetic and potential energy. *Kinetic energy* is energy at work or in motion; that is, moving energy—a car in motion or a rotating shaft—has kinetic energy. In billiards, a player gives the cue ball kinetic energy when she strikes the ball with the cue. As the ball rolls, it exerts kinetic energy. When the ball meets another ball, it transmits its kinetic energy, allowing the next ball to be accelerated. *Potential energy* is stored energy, like the energy stored in a coiled or stretched spring, or an object stationed above a table. A roller coaster has the greatest potential energy when it is stopped at the top of a long drop. Another example of potential energy is when a can of carbonated soda remains unopened. The can is pressurized with gas that is not in motion but that has potential energy. Once the can is opened, the gas is released and the potential energy is converted to kinetic energy.

According to the Conservation Law of Energy, energy cannot be made or destroyed, but can be made to change forms. Moreover, when energy changes from one form to another, the amount of energy stays the same. Let's consider an example of the

Conservation Law of Energy: The energy of something is measured to start with; the energy changes from potential (stored) energy to kinetic (moving) and back again; at the end, the energy is measured again. The energy measured at the start is the same as that measured at the end; it will always be the same. One caveat to this explanation is that we now know that matter can be made into energy through processes modified or amplified to become the Law of Conservation of Matter and Energy.

TYPES OF ENERGY

The main sources of green energy are wind energy, solar power, and hydroelectric power (including pumped hydropower storage systems and tidal energy, which uses ocean energy from the ties in the sea). Both sonar and wind power can be produced on a small scale for people's private use, for business, or alternatively, they can be designed and constructed on a larger, industrial scale.

There are many types of energy. For example:

- Kinetic (motion) energy
- Water energy
- Potential (at rest) energy
- Elastic energy
- Nuclear energy
- Chemical energy
- Sound energy
- Internal energy
- Heat/Thermal energy
- Light (radiant) energy
- Electric energy

Energy sources can also be categorized as renewable or nonrenewable. The Energy Information Administration (EIA, 2009) points out that when we use electricity in our home, the electrical power was probably generated by burning coal, by a nuclear reaction, or by a hydroelectric plant at a dam. Therefore, coal, nuclear, and hydropower are called energy sources. When we fill up a gas tank, the source might be petroleum or ethanol made by growing and processing corn.

As mentioned, energy sources are divided into two groups—*renewable* (an energy source that can be easily replenished; includes Green Energy) and *nonrenewable* (an energy source that we are using up and cannot recreate; petroleum, for example, was formed millions of years ago from the remains of ancient sea plants and animals). In the United States, most of our energy comes from nonrenewable energy sources. Coal, petroleum, natural gas, propane, and uranium are nonrenewable energy sources. They are used to make electricity, to heat our homes, to move our cars, and to manufacture all kinds of products. Green (renewable) and nonrenewable energy sources can be used to produce secondary energy sources including electricity and hydrogen. Renewable/green energy sources include those shown in Figure 1.1.

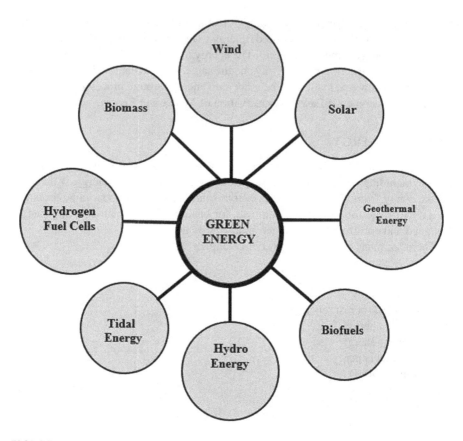

FIGURE 1.1   Renewable green energy sources.

NONRENEWABLE ENERGY VERSUS GREEN ENERGY

Green energy (an energy source that can be easily replenished) is the focus of this text. Unfortunately (depending on your point of view), nonrenewable energy sources on Earth are available in limited quantity and may vanish within the next 100 years. Moreover, keep in mind that nonrenewable sources *are not* environmentally friendly and can have serious effects on our health. Notwithstanding the environmental and health impacts of using nonrenewable energy sources, it is important to point out both sides of the argument, that is, the benefits derived, and the non-benefits obtained from using these sources. For example, nonrenewable energy sources are beneficial in that:

- They are easy to use.
- A small amount of nuclear energy will produce a large amount of power.
- They have little competition.
- They are relatively cheap when converting from one type of energy to another.

Nonrenewable energy sources are not beneficial in that:

- They will expire someday.
- The speed at which such resources are being used can bring about serious environmental changes.
- Nonrenewable sources release toxic gases in the air when burned and can further exacerbate ongoing, cyclical climate change.
- Because nonrenewable sources are becoming scarcer, the prices of these sources will begin to soar.

The benefits of using green energy sources include the following:

- Wind, sun, ocean, and geothermal energy are available in abundant quantities and are free to use.
- Green energy sources have low carbon emissions; therefore, they are considered environmentally friendly.
- Green energy helps stimulate the economy and creates job opportunities.
- Green energy sources enable the country to become energy independent, not having to rely on foreign (often hostile) sources.

Non-benefits of green energy sources include the following:

- Initial set-up costs of renewable energy sources are quite high.
- Solar energy is limited to daytime use and not during the night or the rainy season.
- Geothermal energy can bring toxic chemicals beneath the earth's surface to the top and can create environmental damage.
- Hydroelectric dams are expensive to build and can affect natural flow and wildlife.
- Wind energy production requires high winds and therefore must be sited properly. Also, they are tall structures that can affect bird populations. There have also been incidents whereby wind turbines tipped over or were internally poorly constructed. *Note*: this finding is based on the author's personal experience and studies.

DID YOU KNOW?

The largest consumer of fossil fuels in modern agriculture is ammonia production (for fertilizer) via the Haber process, which is essential to high-yielding intensive agriculture. The specific fossil fuel input to fertilizer production is primarily natural gas, to provide hydrogen via steam reforming.

Use of energy in the United States is shared by four major sectors of the economy. Each end-use sector consumes electricity produced by the electric power sector (EIA, 2023):

- Commercial—18%—includes buildings such as offices, malls, stores, schools, hospitals, hotels, warehouses, restaurants, places of worship, and more.
- Industrial—33%—includes facilities and equipment used for manufacturing, agriculture, mining, and construction.
- Residential—21%—consists of homes and apartments.
- Transportation—28%—comprises vehicles that transport people or goods, such as: cars, trucks, buses, motorcycles, trains, subways, aircraft, boats, barges, and even hot air balloons.

Primary energy consumption in the United States was almost three times greater in 2012 as in 1949. In all but 18 of the years between 1949 and 2012, primary energy consumption increased over the previous year.

The year 2009 provided a sharp contrast to the historical trend, in part due to the economic recession. Real gross domestic product (GDP) fell 2% compared to 2008, and energy consumption declined by nearly 5%, the largest single-year decline since 1949. Decreases occurred in all four major end-use sectors: residential—3%, commercial—3%, industrial—9%, and transportation—3% (EIA, 2023).

GREEN ENERGY FORMS

The forms of green energy include the following, as shown in Figure 1.1 and listed here:

- Solar
- Hydro
- Wind
- Geothermal
- Biomass
- Ocean/Marine Power (tidal power)
- Biofuels
- Hydrogen fuel cells

THE BOTTOM LINE

Green energy production and use is on the increase because it's cheaper, is much better for our health, is great for the environment, and in this author's opinion, it is the future.

NOTE

1 Much of the information in this section is from F.R. Spellman (2014).*The Environmental Impacts of Renewable Energy*. Boca Raton: CRC Press.

REFERENCES AND RECOMMENDED READING

Brandt, A.R. 2007. Testing Hubbert. *Energy Policy* **35**(5): 3074–3088.

Cambridge Energy Research Associates (CERA). 2006. CERA says peak oil theory is faulty. Energy Bulletin (Cambridge Energy Research Associates). Accessed 01/16/23 @ www. energybulletinnet/22381.html

Campbell, C.J. and Laherrere, J.H. 1998. The end of cheap oil. *Scientific American*, March, **278**: 78–83.

Deffeyes, K.S. 2007. *Current Events—Join Us as We Watch the Crisis Unfolding*. Princeton University: Beyond Oil. Accessed 01/16/23 @ www.princeton.edu/hubbert/current-eve nts.html

Energy Information Administration (EIA). 2009. *What Is Energy? Explained*. US Energy Information Administration. Accessed 01/22/23 @ http://tonto.eia.doe.gov/energyex plained/print.cfm?page=about_sources_of_energy

Energy Information Administration (EIA). 2023. *Use of Energy in the United States Explained*. US Energy Information Administration. Accessed 02/10/23 @ www.eia.gov/energyex plianed/index.cfm?

Gwyn, R. 2004. *Demand for Oil Outstripping Supply*. Accessed 1/15/23 @ www.commondre ams.org/views04/0128-10.htm

Hirsch, R.L. 2005. *The Mitigation of the Peaking of World Oil Production Summary of an Analysis*. Accessed 01/16/23 @ www.peakoil.net/USDOE.html

Hubbert, M.K. 1956. *Nuclear Energy and the Fossil Fuels "Drilling and Production Practice."* Accessed 01/08/23 @ www.hubbertpeak.com/hubbert/1956/1956.pdf

Koppelaar, R.H.E.M. 2006. *World Production and Peaking Outlook*. Accessed 01/12/23 @ http://peakoil.nl/wp-content/upoads/2006/09/aspnl_2005_report.pdf

Rich, A. 1973. *Diving into the Wreck*. Accessed 8/9/23 @ www.poetrysoup.com/famous/ poems/diving_into_the_wrreck_9335

Shakir, U. 2023. *EV Charging in the US Is Still a No-Good Bad Time—and Somehow Getting Worse*. Accessed 8/16/23 @ The Verge. www.theverge.com/about-the-verge

US Department of Energy (USDOE). 2005. Peaking of World Oil Production: Impacts, Mitigation, & Risk Management. Accessed 01/05/23 @ www.netl.doe.gov/publications. others/pdf/Oil_Peaking_NETL.pdf

US Department of Energy (USDOE). 2007. *Peak Oil—The Turning Point*. Washington, DC: US Department of Energy.

2 Solar Energy

The Sun, with all the planets
revolving around it, and depending
on it, can still ripen a bunch of
grapes as though it had nothing
else in the universe to do ...

–Galileo Galilei[1]

An important quote from one of the only true Americans:
The Sun, the darkness ...
They are all listening to what we have to say.

–Geronimo[2]

INTRODUCTION

It is fitting to begin our discussion of the various kinds of renewable energy with the sun—the medium-sized star that symbolizes life, power, strength, force, clarity, and, yes, energy. The sun nourishes our planet. Solar energy is a green/renewable resource because it is continuously supplied to the Earth by the sun. When we consider the sun and solar energy first, we quickly realize that there is nothing new about green energy. The sun was the first energy source; it has been around for at least 4.5 billion years if anything else we are familiar with. On Earth, without the sun there is nothing—absolutely nothing, zero. There is the old 5-5-5-thumb rule that guides the essence of life on Earth; that is, we commonly state that one cannot survive a lack of breath (oxygen) after five minutes; we must have water available to drink in five or fewer days; and we need food to eat in five weeks or less to stay alive. All this is true, of course, but without the sun the 5-5-5-rule has no validity—without the sun nothing related to life has validity—without the sun there is nothing: no air, no water, no food and no life.

The sun, the ultimate energy source, provided light and heat to the first humans. During daylight, people searched for food. They hunted and gathered and probably stayed together for their safety. When nightfall arrived, we can only imagine that

DOI: 10.1201/9781003439059-3

they huddled together for warmth and reassurance considering the stars and moon and campfires only, waiting for the sun and its live-giving and life-sustaining light to return.

Solar energy (a term used interchangeably with solar power) uses the power of the sun, employing various technologies, "directly and/or indirectly," to produce energy. Solar energy is one of the best green energy sources available because it is one the cleanest sources of energy. Direct solar radiation absorbed in solar collectors can provide space heating and hot water. Passive solar can be used to enhance the solar energy use in buildings for space heating and lighting requirements. Solar energy can also be used to produce electricity, and this is the green energy resources that is the focus attention in this section.

Radiant energy from the sun, in the form of photons, strikes the surface of the Earth with the average equivalent of about 168 kWh of energy (equal to 575,000 Btus thermal energy) per square foot per year; it varies, of course, with location, cloud cover, and orientation with the surface (Hanson, 2004). The question then becomes, how much of this energy is used in the United States by various consumers? In attempting to answer this question, Table 2.1 highlights solar energy's quadrillion Btu ranking in current renewable energy source use. As indicated in the table, the energy consumption by energy source computations is from the year 2022. Thus, the 12% solar/PV quadrillion Btu figure is expected to steadily climb to an increasingly higher level and this should be reflected in the 2022–2023 figures when they are released.

In its Solar Energy Technology Program, the USDOE (2009a) points out that not only do solar technologies diversify the energy supply but they also reduce the country's dependence on imported fuel. Moreover, solar energy technologies provide a beneficial environmental impact by helping improve air quality and offset greenhouse gas emissions. An additional (and significant) benefit of a growing solar technology industry is that it stimulates our economy by creating jobs in solar manufacturing and installation. According to the USDOE (2009b), the two solar electric

---

TABLE 2.1
US Energy Consumption by Energy Source, 2022 (Quadrillion Btu & %)

Energy Source	2022
Total	**97.33**
Renewable	12.16
Biomass (biofuels, waste, wood and wood-derived)	40%
Biofuels	19%
Waste	4%
Wood derived Fuels	17%
Geothermal	2%
Hydroelectric Conventional	19%
Solar/PV	**12%**
Wind	27%

Source: EIA 2022. *US Energy consumption by Energy Source.* Accessed 06/25/23 @ www.eia.energyex plained/us-energy-facts.

technologies with the greatest potential (based on cost-effectiveness) are concentrated solar power (CSP) and photovoltaics (PV).

CONCENTRATED SOLAR POWER (CSP)[3]

Concentrated solar power (CSP), also known as concentrating solar power, concentrated solar thermal) offers a utility-scale, reliable, firm, dispatchable renewable energy option that can help meet a nation's demand for electricity. The Office of Energy Efficiency & Renewable Energy (EERE, 2008a) points out that nine trough plants producing more than 400 megawatts (MW) of electricity have been operating reliably in the California Mojave Desert since the 1980s. CSP plants produce power by first using mirrors to focus sunlight to heat a working fluid, a thermal receiver, like a boiler tube. The receiver absorbs and converts sunlight into heat. Ultimately, this high-temperature fluid is used to spin a turbine or power an engine that drives a generator which produces electricity.

Concentrating solar power systems can be classified by how they collect solar energy, by using linear concentrators, dish/engines, or power tower systems (NREL, 2009). All three are typically engineered with tracking devices for following the sun both seasonally and throughout the day in order to maximize the system's electrical output (Chiras, 2002).

LINEAR CONCENTRATORS

As mentioned, linear concentrating solar power (CSP) concentrators or collectors are one of the three types of CSP systems in used today. Linear CSP collectors capture or collect the sun's energy using long rectangular, curved (U-shaped) mirrors. The mirrors are tilted toward the sun, focusing sunlight on tubes (or receivers) that run the length of the mirrors. The reflected sunlight heats a fluid flowing through the tubes. The hot fluid is then used to boil water to create superheated steam that spins a conventional steam-turbine generator to produce electricity. Alternatively, steam can be generated directly in the solar field, eliminating the need for costly heat exchangers.

Linear concentrating collector fields consist of many collectors in parallel rows that are typically aligned in a north-south orientation to maximize both annual and summertime energy collection. With a single-axis sun-tracking system, this configuration enables the mirrors to track the sun from east to west during the day, ensuring that the sun reflects continuously onto the receiver tubes (EERE, 2008b).

There are two major types of linear concentrator systems: parabolic trough systems, where receiver tubes are positioned along the focal line of each parabolic mirror, and linear Fresnel reflector systems, where one receiver tube is positioned above several mirrors to allow the mirrors greater mobility in tracking the sun.

PARABOLIC TROUGH SYSTEMS

EERE (2008b) points out that the predominant CSP systems currently in operation in the United States are linear concentrators using parabolic trough collectors. All rays of light that enter parallel to the axis of a parabolic-shaped mirror will be reflected to one point, the focus. Hence, it is possible to concentrate virtually all the radiation incidents upon the mirror in a relatively small area in the focus (Perez-Blanco, 2009). Parabolic trough systems form a long-curved array of mirrors (usually coated silver or polished aluminum). Trough systems have been developed in the United States, Spain, and Japan (Chiras, 2002). The receiver tube (Dewar tube) runs its length and is positioned along the focal line of each parabola-shaped reflector (Duffie and Beckman, 1991; Patel, 1999). The tube is fixed to the mirror structure and the heated fluid—either a heat-transfer fluid (usually oil) or water/stream—flows through and out of the field of solar mirrors to where it is used to create steam (or, for the case of a water/steam receiver, it is sent directly to the turbine (i.e., the turbine is the prime mover of the electrical generator that produces electrical current flow in an accompanying electrical distribution system). Temperatures in these systems range from 150° to 750° F (80°–400° C).

Currently, the largest individual trough systems generate 80 megawatts of electricity. However, individual systems being developed will generate 250 megawatts. In addition, individual systems can be collocated in power parks. This capacity would be constrained only by transmission capacity and availability of contiguous land area.

Trough designs can incorporate thermal storage. Thermal storage systems are discussed later, but for informative purposes, we can say that in such a system, the collector field is oversized to heat a storage system during the day that can be used in the evening or during cloudy weather to generate additional steam to produce electricity. Parabolic trough plants can be designed as hybrids, meaning that they use fossil fuel to supplement the solar output during periods of low solar radiation. In such a design, a natural-gas-fired heater or gas-steam boiler/reheater is used. In the future, troughs may be integrated with existing or new combined-cycle natural-gas and coal-fired plants (EERE, 2008b).

FRESNEL REFLECTOR SYSTEMS

A second linear concentrator technology, a solar power collector, is the linear Fresnel reflector system. Flat or slightly curved mirrors mounted on trackers on the ground are configured to reflect sunlight onto a water-filled receiver tube fixed in space above these mirrors. A small parabolic mirror is sometimes added atop the receiver to further focus the sunlight. The key advantage of Fresnel reflectors systems is their simplicity as compared to other systems.

DISH/ENGINE SYSTEM

Dish/engine systems use a mirrored dish-shaped parabolic mirror like a very large satellite dish. The dish-shaped surface directs and concentrates sunlight onto a thermal receiver (the interface between the dish and the engine/generator), which absorbs and collects the heat and transfers it to the engine generator. Solar dish/engine systems convert the energy from the sun into electricity at a very high efficiency. The power conversion unit includes the thermal receiver and the engine/generator. It absorbs the concentrated beams of solar energy, converts them to heat, and transfers the heat to the engine/generator. A thermal receiver can be a bank of tubes with a cooling fluid—usually hydrogen or helium—that typically is the heat-transfer medium and the working fluid for an engine. Alternate thermal receivers are heat pipes, where the boiling and condensing of an intermediate fluid transfers the heat to the engine. The most common type of heat engine used today in dish/engine systems is the Stirling engine (conceived in 1816 and mentioned earlier). This system uses the fluid heated by the receiver to move pistons and create mechanical power. The mechanical power is then used to run a generator or alternator to produce electricity (USDOE, 1998).

Dish/engine systems use dual-axis collectors to track the sun. As mentioned, the ideal concentrator shape is parabolic, created with by a single reflective surface or multiple reflector, or facets. Many options exist for receiver and engine type, including, as mentioned, the Stirling cycle, microturbine, and concentration photovoltaic modules.

Because of the high concentration ratios achievable with parabolic dishes, and the small size of the receiver, solar dishes are efficient (as high as 30% efficient) at collected solar energy at very high temperatures (900°–2700° F; 500°–1500° C).

The USDOE (1998) points out that solar dish/engine systems have environmental, operational, and potential economic advantages over more conventional power generation options because they:

- produce zero emissions when operating on solar energy
- operate more quietly than diesel or gasoline engines
- are easier to operate and maintain than conventional engines
- start up and shut down automatically
- operate for long periods with minimal maintenance

Solar dish/engine systems are well suited to and often used for nontraditional power generation because of their size and durability. Individual units range in size from 10 kilowatts to 25 kilowatts. They can operate independently of power grids in remote sunny locations for uses such as pumping water and providing power to people living in isolated locations.

As mentioned, dish/engine systems produce high temperatures, which in turn can be used to produce steam that can be used either for electricity production or select high-temperature industrial processes (Chiras, 2002). Dish/engine systems also can be linked together to provide utility-scale power to a transmission grid. Such systems could be located near consumers, substantially reducing the need for building or upgrading transmission capacity. Largely because of their high efficiency, the cost of these systems is expected to be lower than that of other solar systems for these applications (USDOE, 1998).

POWER TOWER SYSTEM

The power tower or the central receiver system consists of a large field of flat, sun-tracking mirrors known as heliostats (silver-laminated acrylic membranes) focused to concentrate sunlight onto a receiver on the top of a tower. A heat-transfer fluid heated in the receiver is used to generate steam, which, in turn, is used in a conventional turbine generator to produce electricity. Some power towers use water/steam as the heat-transfer fluid. Other advanced designs are experimenting with molten (liquefied) nitrate

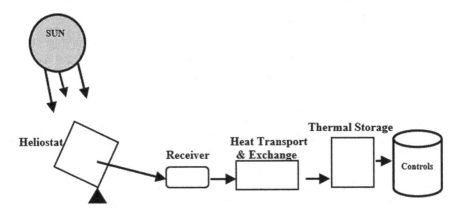

FIGURE 2.1   Solar power tower main components.

salt because of its superior heat-transfer and energy-storage (heat retention) capabilities, allowing continued electricity production for several consecutive cloudy days (NREL, 2010). The energy-storage capability, or thermal storage, allows the system to continue to dispatch electricity during cloudy weather or at night. Therefore, a power tower system is composed of five main components: heliostats, receiver, heat transport and exchange, thermal storage, and controls (see Figure 2.1). Individual commercial plants can be sized to produce up to 200 megawatts (and more) of electricity.

DID YOU KNOW?

Smaller CSP systems can be located directly where the power is needed. For example, a single, dish/engine system can produce 3 to 25 kilowatts of power and is well suited for such distributed applications. Larger, utility-scale CSP applications provide hundreds of megawatts of electricity for the power grid. Both linear concentrator and power tower systems can be easily integrated with thermal storage, helping generate electricity during cloudy periods or at night. Alternatively, these systems can be combined with natural gas, and the resulting hybrid power plants can provide high-value, dispatchable power throughout the day.

DID YOU KNOW?

Concentrated solar power plants use water not only for lens cleaning but also for stream cycle production. Water consumption is an issue because these plants need water and they are most effective in locations where the sun is most intense, which in turn often corresponds to places like the Mohave Desert where there is little water (USDOE, 2008).

THERMAL STORAGE (EERE, 2008C)

Thermal storage (TES) has become a critical aspect of any concentrating solar power (CSP) system deployed today. One challenge facing the widespread use of solar energy is the reduced or curtailed energy production when the sun sets or is blocked by clouds. Thermal energy storage provides a workable solution to this challenge. In a CSP system, as mentioned, the sun's rays are reflected onto a receiver, creating heat that is then used to generate electricity. If the receiver contains oil or molten salt as the heat transfer medium, then the thermal energy can be stored for later use. This allows CSP systems to be cost-competitive options for providing clean, renewable energy. Presently, steam-based receivers cannot store thermal energy for later use.

An important criterion in selecting a storage medium, such as the oil or molten salt referred to above, is the *specific heat* of the substance. Specific heat (c) is the measure of the heat energy required to increase the temperature of a unit quantity of a substance by a unit of temperature. For example, at a temperature of 15° C, the heat energy required to raise water's temperature one kelvin (equal to one degree Celsius) is 4,186 joule per kilogram. Different materials absorb different amounts of heat in undergoing the same temperature increase. The relationship between a temperature change ΔT and the amount of heat Q added (or subtracted) is given by

$$Q = mc\ \Delta T, \qquad (2.1)$$

where c is the specific heat and m the mass of the substance. Table 2.2 lists some specific heats for different materials per pound and per cubic foot. Another important parameter related to specific heat and thermal energy storage materials is *heat capacity*, which is defined as the ratio of the heat energy absorbed by a substance to the substance's increase in temperature and is given by

$$\text{Heat Capacity} = \text{Specific Heat x Density} \qquad (2.2)$$

Another medium for thermal storage are *phase change materials*, which are classified as latent heat storage units. Specifically, a phase material is a substance with a high heat of fusion (i.e., the amount of thermal energy that must be absorbed or evolved for one mole of a substance to change states from a solid to a liquid or vice versa without a temperature change), which, melting and solidifying at a certain temperature, can store and release large amounts of energy. Heat is absorbed or released when the material changes from solid to liquid and vice versa. One common group of substances being used in as phase change materials for solar applications (and elsewhere) are eutectic salts; salts are combined with water such as sodium sulfate decahydrate, also known as Glauber's Salt. This salt melts at 91° F with the addition of 108 Btu per pound. Conversely, when the temperature drops below 91° F, 108 Btu per pound of heat energy is released as the salt solidifies (Hinrichs & Kleinbach, 2006).

Several TES technologies have been tested and implemented since 1985. These include the two-tank direct system, the two-tank indirect system, and the single-tank thermocline system.

TABLE 2.2
Thermal Energy Storage Materials

Material	Specific Heat Btu/lb/°F	Density (lb/ft³)	kg/m³	Heat Capacity (Btu/ft³-°F)	(kj/m³-°C)
Water	1.00	62	1000	62	4186
Iron	0.12	490	7860	59	3521
Copper	0.09	555	8920	50	3420
Aluminum	0.22	170	2700	37	2430
Concrete	0.23	140	2250	32	2160
Stone	0.21	170	2700	36	2270
White Pine	0.67	27	435	18	1220
Sand	0.19	95	1530	18	1540
Air	0.24	0.075	1.29	0.02	1.3

TWO-TANK DIRECT SYSTEM

Solar thermal energy in this system is stored in the same fluid used to collect it. The fluid is stored in two tanks—one at high temperature and the other at low temperature. Fluid from the low-temperature tank flows through the solar collector or receiver, where solar energy heats it to the high temperature and it then flows back to the high-temperature tank for storage. Fluid from the high-temperature tank flows through a heat exchanger, where it generates steam for electricity production. The fluid exits the heat exchanger at the low temperature and returns to the low temperature tank. Two-tank direct storage was used in early parabolic trough power plants and at the Solar Two power tower in California. The trough plants used mineral oil as the heat-transfer and storage fluid; Solar Two used molten salt.

TWO-TANK INDIRECT SYSTEM

This system functions in the same way as the two-tank direct system, except different fluids are used as the heat-transfer and storage fluids. This system is used in plants where the heat-transfer fluid is too expensive or not suited for use as the storage fluid. The storage fluid from the low-temperature tank flows through an extra heat exchanger, where it is heated by the high-temperature heat-transfer fluid. The high-temperature storage fluid then flows back to the high-temperature storage tank. The fluid exits this heat exchanger at a low temperature and returns to the solar collector or receiver, where it is heated back to the high temperature. Storage fluid from the high-temperature tank is used to generate steam in the same manner as the two-tank direct system. The indirect system requires an extra heat exchanger, which adds cost to the system. This system will be used in many of the parabolic power plants in Spain and has also been proposed for several US parabolic plants. The plants will use organic oil as the heat-transfer fluid and motor salt as the storage fluid.

SINGLE-TANK THERMOCLINE SYSTEM

This system stores thermal energy in a solid medium—most commonly silica sand—located in a single tank. At any time during operation, a portion of the medium is at high temperature and a portion is at lower temperature. The hot- and cold-temperature regions are separated by a temperature gradient or thermocline (see Figure 2.2). High-temperature heat-transfer fluid flows into the top of the thermocline and exits the bottom at low temperature. This process moves the thermocline downward and adds thermal energy to the system for stage. Reversing the flow moves the thermocline upward and removes thermal energy from the system to generate steam and electricity. Buoyancy effects create thermal stratification of the fluid within the tank, which helps stabilize and maintain the thermocline.

PHOTOVOLTAICS (PV)

Hinrichs and Kleinbach (2006) point out that Photovoltaic (PV) has been and will continue to be one of the more glamorous technologies in the energy field. Photovoltaic (Gr. *photo* light, and *volt*, electricity pioneer Alessandro Volta) technology makes use of the abundant energy in the sun, and it has little impact on our environment. *Photovoltaics* is the direct conversion of light (photons) into electricity (voltage) at the atomic level. Some materials exhibit a property known as the *photoelectric effect* (discovered and described by Becquerel in 1839) that causes them to absorb photons of light and release electrons. When these free electrons are captured, an electric

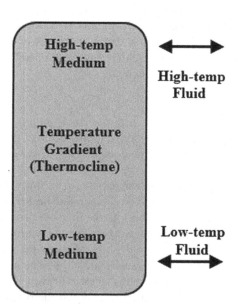

FIGURE 2.2 Single-tank thermocline thermal energy storage system.

Note: Using a solid storage medium and only needing one tank reduces the cost of the single-tank thermocline system relative to the two-tank systems.

current results (i.e., electricity is the flow of free electrons) that can be used as electricity. The first photovoltaic module (billed as a solar battery) was built by Bell laboratories in 1954. In the 1960s, the US space program began to make the first serious use of the technology to provide power aboard spacecraft. Space program use helped this technology make giant advancements in reliability and helped lower cost. However, it was the oil embargo of the 1970s (the so-called energy crisis) that propelled photovoltaic technology to the forefront of recognition for use other than space applications only. Photovoltaics can be used in a wide range of products, from small consumer items to large commercial solar electric systems.

Figure 2.3 illustrates photoelectric effect when light shines on the negative plate, electrons are emitted with an amount of kinetic energy inversely proportional to the wavelength of the incident light. Figure 2.4 illustrates the operation of a basic *photovoltaic cell*, also called a *solar cell*. Solar cells are made of silicon, and other semiconductor materials such as germanium, gallium arsenide, and silicon carbide are used in the microelectronics industry. For solar cells, a thin semiconductor wafer is specially treated to form an electric field, positive on one side and negative on the other. When light energy strikes the solar cell, electrons are jarred loose from the atoms in the semiconductor material (see Figures 2.3, 2.4). If electrical conductors are attached to the positive and negative sides, forming an electrical circuit, the electrons can be captured in the form of an electrical current—that is, again, recall that electron flow is electricity. This electricity can then be used to power a load, such as a light, a tool, a toaster, and other electrical appliances and/or apparatuses.

Several solar cells electrically connected to each other and mounted on a support panel or frame is called a photovoltaic module (see Figure 2.5). Solar panels used to power homes and businesses are typically made from solar cells combined into modules that hold about 40 cells. Modules are designed to supply electricity at a certain voltage, such as a common 12-volt system. The current produced is directly dependent on how much light strikes the module.

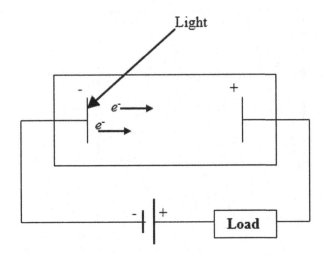

FIGURE 2.3 Photoelectric effect.

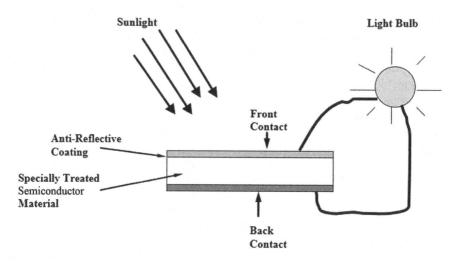

FIGURE 2.4 Operation of basic photovoltaic cell.

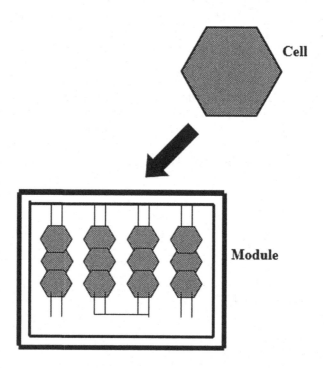

FIGURE 2.5 Single solar cell and solar cell module.

Multiple modules can be wired together to form an array. In general, the larger the area of a module or array, the more electricity that will be produced. Photovoltaic modules and arrays produce direct-current (dc) electricity. They can be connected in both series and parallel electrical arrangements to produce any required voltage and current combination.

SOLAR ENERGY AND DOMESTIC/INDUSTRIAL USE

When thinking about a potential renewable energy source, we first want to know if it is reliable, clean, and affordable. This is especially the case for consumers and business owners contemplating solar applications for both home and business use. Until recently, solar energy has been used in both the home and industrial locations but only on a somewhat limited basis. This trend is changing, especially in those locations where sunlight is prevalent throughout the year. With the soaring costs of fossil fuel supplies and their pending decreased availability, solar power is beginning to receive more attention. For domestic use, solar technologies: photovoltaics, passive heating, daylighting, and water heating; change to are currently in development. Industrial and commercial facilities may use the same solar technologies that are used for residential buildings. These nonresidential buildings can also use solar energy technologies that would be impractical for a home. These technologies include ventilation air preheating, solar process heating, and solar cooling.

SOLAR HOT WATER

Just as the sun heats the surface layers of exposed bodies of water—ponds, lakes, streams, and oceans—it can also heat water used in buildings and swimming pools. To harness the sun's ability to heat water, two main parts are added to the process: a solar collector and a storage tank. The most common collector is called a *flat-plate collector* (see Figure 2.4). The flat-plate collector system is the most used collector and is ideal for applications that require water temperatures under 140° F (60° C). Mounted on the roof, it consists of a thin, flat, rectangular, insulated box with a transparent-glass cover that faces the sun. Small copper tubes run through the box and carry the fluid—either water or other fluid, such as an antifreeze solution—to be heated. The tubes, typically arranged in series (i.e., water flows in one end and out the other end) are attached to an absorber plate, which, along with the tubes, is painted black to absorb the heat. As heat builds up in the collector, it heats the fluid (water or propylene glycol) passing through the tubes. The motive force for passing fluid through the tubes can be accomplished either by active or passive means. In an active system, the most common type of system, water heaters rely on an electric pump and controller to circulate water, or other heat-transfer fluids through the collectors. Passive solar water-heater systems rely on gravity and the tendency for water to naturally circulate as it is heated. Because passive systems have no "moving parts," they are more reliable, easier to maintain, and often have a longer life than active systems.

The storage tank holds the hot liquid (see Figure 2.6). It is usually a large well-insulated tank, but modified water heaters can also be used to store the hot liquid.

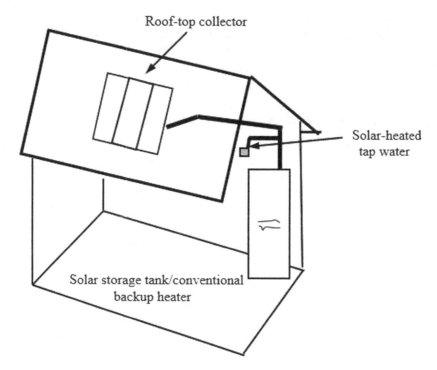

FIGURE 2.6   Simplified representation of a home solar water-heating system.

When the fluid used is other than water, the water is heated by passing it through a coil of tubing in the tank, which is full of hot fluid. Active solar systems usually include a storage tank along with a conventional water heater. In two-tank systems, the solar water heater preheats water before it enters the conventional water heater.

SOLAR PROCESS HEAT

In addition to conventional domestic water heating systems, commercial and industrial buildings may use the same solar technologies—photovoltaics, passive heating, daylighting, and water heating. These nonresidential buildings can also use solar energy technologies that would be impractical for a home. These technologies include ventilation air preheating, solar process heating, and solar space cooling

Space Heating

In order to maintain indoor air quality (IAQ), many large buildings need ventilated air. Although all that fresh air is great for IAQ, in cold climates, heating this air can use large amounts of energy and can be very expensive. But an elegantly simple solar ventilation system can preheat the air, saving both energy and money. This type of system typically uses a transpired collector.

Transpired air collector systems essentially consist of a dark-colored, perforated facade installed on a building's south-facing wall. An added fan or the building's existing ventilation system draws ventilation air into the building through the perforated absorber plate on the facade and up the plenum (the air space between the absorber and the south wall). According to the USDOE (2006), solar energy absorbed by the dark absorber and transferred to the air flowing through it can preheat the intake air by as much as 40° F (22° C). Reduced heating costs will pay for the systems in 3–12 years.

In addition to meeting a portion of a building's heating load with clean, free solar energy, the transpired collector helps save energy and money in other ways. It recaptures heat loss through a building's south-facing wall; heat that escapes through the south wall is captured in the air space between the structural wall and the transpired collector and returned to the interior. Also, by introducing make-up air through ceiling-mounted ducts, the system eliminates the wasteful air stratification that often plagues high-ceiling buildings (USDOE, 2006).

To illustrate an active, real-world example of transpired collectors (used as solar preheaters for outdoor ventilation air), it is probably best to use an actual example of a working system in an industrial application. The following case study is such an example; it has been monitored extensively by experts.

CASE STUDY—GENERAL MOTORS BATTERY PLANT[4]

The General Motors battery plant in Oshawa, Canada, is a 100,000-foot^2 facility in which automotive batteries are manufactured. The plant was built in the 1970s and consists of an open shop floor and a 298-foot-high ceiling. GM operates two full-time production shifts within the plant and conducts maintenance activities at night and on weekends, so the building is continuously occupied.

DID YOU KNOW?

Besides the questions potential electric car purchasers have related to range anxiety, lack of charging availability, and the cost of electric vehicles (EVs), there are two more important and pressing questions that potential customers have—having caused hesitation in customer purchases: First, how long do EV batteries last? Second, how much does it cost to replace a spent EV battery? OK, the truth be told, we do not know how long EV batteries last. This is case because EVs are relatively new with original batteries. And evolving technology … hmmm, about a six-year record to date. Thus, it is impossible to make a definitive judgment on how long EV batteries will or can last. The EV battery is its most expensive component (at the present time), costing maybe as much as $22,000 to replace a Tesla battery. This is important to purchasers, but also how long will the power last before being spent is the other pressing question. How long of a warranty is needed? Are the batteries covered for three or four or five years, or longer? Do these questions even matter? Are we moving toward EV technology change too soon? Does anyone know? How about hybrid vehicles? Are they a better option, or choice?

Until the early 1990s, GM relied solely on a team-operated fan coil system for space heating, but the system was incapable of providing the necessary quantities of heated outdoor air. As a result, the plant was not being adequately ventilated.

In 1991, plant management installed a transpired collector to correct the ventilation problems. During the following two years the transpired collector system was modified slightly to improve airflow; the original fans and motors were replaced with axial fans and high-efficiency motors, and the original ducting was replaced with the upgraded fabric ducting.

The GM plant collector comprises 4,520 feet^2 of absorber sheeting. The lower 21 feet of the transpired collector is black, perforated, aluminum wall cladding with 1.6-mm holes totaling 2% porosity. The average depth of the plenum between the transpired collector and the plant's structural wall is six inches. The canopy at the top of the wall acts as both a manifold for the airflow and a solar heat collection device. The canopy face is made of perforated plate with 1% porosity. The transpired collector covers about 50% of the total area of the plant's south-facing wall; the remainder of the south facade has shipping doors and other obstructions that make it unsuitable for mounting collector cladding.

The GM transpired collector has fan/distribution systems, each consisting of a constant-speed fan, a recirculation damper system, and a fabric distribution duct. The total airflow delivered by the system's fans is 40,000 cfm. Both recirculated air and air drawn through the solar collectors make up this flow; the percentages of each depend on the temperature of the air coming from the collector.

The GM battery plant's transpired collector has been monitored extensively since it was installed. The data in this case study reflect the performance of the system during the 1993/94 heating season. An in-depth report on the monitoring program is available (Enermodal Engineering Ltd., 1995).

The data shows that the annual energy savings for the 4,520-foot^2 collector was 940 million Btu/year: 678 MBtu resulted from the thermal energy gained directly from the outside air as it passed through the absorber, and 262 MBtu resulted from heat loss recaptured by the wall form inside the building. Other possible energy-saving mechanisms—such as destratification and heat recapture—likely contributed to improved system performance; however, these effects are highly structure-specific and have not been incorporated into the savings reported here.

The cost of the transpired collector system at the GM plant was $66,530, or $14.72/ft^2 of installed collector. The cost per square foot is higher than typical installations for two reasons: 1) this system was installed soon after the technology was introduced, before design and installation procedures had been streamlined, and 2) the cost includes the fan and ducting modifications that were implemented during the first two years of operation.

It is important to point out that the GM transpired collector had several operational problems. After the system was initially installed, employees complained about fan noise and cold drafts, and they occasionally disabled the system. The fan and duct upgrades described previously eliminated the problems on one of the fan systems; the other fan continues to generate noise, and employees still disable it when working in the immediate vicinity. The manufacturer (Conserval) has addressed these complaints by specifying smaller, but more numerous, fans in subsequent installations.

Also, both bypass dampers and a recirculation damper required additional maintenance. The recirculation damper became stuck in the full recirculation mode, and a new modulating motor was installed to fix the problem. The bypass dampers occasionally became bound, which led to unacceptably high leakage rates. These dampers were kept closed manually through the 1993–1994 heating season.

WATER HEATING

Solar water-heating systems are designed to provide large quantities of hot water for nonresidential buildings. A typical system includes solar collectors that work along with a pump, heat exchanger, and/or one or more large storage tanks. The two main types of solar collectors used for nonresidential buildings—an evacuated-tube collector and a linear concentrator—can operate at high temperatures with high efficiency. An evacuated-tube collector is a set of many double-walled, glass tubes and reflectors to heat the fluid inside the tubes. A vacuum between the two walls insulates the inner tube, retaining the heat. Linear concentrators use long, rectangular, curved (U-shaped) mirrors tilted to focus sunlight on tubes that run along the length of the mirrors. The concentrated sunlight heats the fluid within the tubes.

SPACE COOLING

Space cooling can be accomplished using thermally activated cooling systems (TACS) driven by solar energy. Because of a high initial cost, TACS are not widespread. The two systems currently in operation are solar absorption systems and solar desiccant systems. Solar absorption systems use thermal energy to evaporate a refrigerant fluid to cool the air. In contrast, solar desiccant systems use thermal energy to regenerate desiccants that dry the air, thereby cooling the air. These systems also work well with evaporative coolers in more humid climates.

LET THERE BE NATURAL LIGHT[5]

Lighting is an essential element of any building. Proper lighting improves the aesthetics of indoor spaces and provides illumination for tasks and activities. An efficient lighting strategy, including natural daylighting, can provide proper levels of illumination and reduce energy costs. *Daylighting* is a passive solar design used to illuminate a living space or an industrial or commercial working space. In the Northern Hemisphere, southern exposure receives the greatest amount of sunlight, while the opposite is true in the Southern Hemisphere. In other words, large picture-type windows placed in the south-facing wall or roof allow natural light to penetrate and save on the use of expensive electrical lighting.

Because the proper use of daylighting requires integration of natural and artificial lighting sources early in the building design process, it is important to understand the fundamentals of lighting and lighting technology. Accordingly, in this section, we first present a discussion of lighting technology before discussing the principles and practices associated with daylighting.

LIGHTING TECHNOLOGIES

Lighting technologies include:

- Lamps—lighting sources, like fluorescent and incandescent light bulbs, and solid-state lighting.
- Ballasts—used with electric discharge lamps such as fluorescent lamps, ballasts transform and control electrical power to the light.
- Luminaries (Fixtures)—complete lighting units that contain the bulbs and, if necessary, the ballasts.
- Lighting controls—devices such as timers and sensors that can save energy by turning lights off when not needed.
- Daylighting—the use of natural light in a building.
- Solid-state lighting—no other technology offers as much potential to save energy and enhance the quality of building environments, contributing to energy and climate change solutions.

LAMPS

Commonly called light bulbs, lamps produce light. When comparing lamps, it is important to understand the following performance characteristics:

- Color Rendering Index (CRI)—a measurement of a light source's accuracy in rendering different colors when compared to a reference light source with the same correlated color temperature. The highest attainable CRI is 100. Lamps with CRIs above 70 are typically used in office and living environments.
- Correlated Color Temperature (CCT)—a measurement on the Kelvin (K) scale that indicates the warmth or coolness of a lamp's color appearance. The higher the color temperature, the cooler the color appearance. Typically, a CT rating below 3200 K is considered warm, while a rating above 4000 K is considered cool.
- Efficacy—the ratio of the light output to the power, measured in lumens per watt (lm/w). The higher the efficacy, the more efficient the lamp.

Lamp types include:

Incandescent Lamps—a standard incandescent lamp consists of a large, thin, frosted glass envelope. Inside the glass is an inert gas such as argon and/or nitrogen. At the center of the lamp is a tungsten filament. Electricity heats the filament.

The heated tungsten emits visible light in a process called incandescence. Most standard light bulbs are incandescent lamps. They have a CRI of 100 and CCTs between 2600–3000, making them attractive lighting sources of many applications. However, these bulbs are typically inefficient (10–15 lm/w), converting only about 10% of the energy into light while transforming the rest into heat. Another type of incandescent lamp is the halogen lamp. Halogen lamps also have a CRI of 100. But they are slightly more energy efficient, and they maintain their light output over time. A halogen lamp also uses a tungsten filament. However, the filament is encased inside a much smaller quartz envelope. The gas inside the envelope is from the halogen group. If the temperature is high enough, the halogen gas will combine with tungsten atoms as they evaporate and redeposit them on the filament. This recycling process lets the filament last much longer. In addition, it is now possible to run the filament hotter. This means more light per unit of energy is obtained. Because the quartz envelope is so close to the filament, it becomes about four times hotter than a standard incandescent lamp. As a result of this wasted heat energy, halogen lamps—popular in torchères (i.e., portable electric lamp with a reflector bowl that directs light upward to give indirect illumination)—are not very energy efficient. The exposed heat from halogen torchieres can also pose a serious fire risk, especially near flammable objects. Because of this inefficiency and risk, manufacturers have developed torchieres that can be used in other lamps, such as compact fluorescent lamps.

Fluorescent Lamps—(70–100 lm/w) consist of a sealed glass tube. The tube contains a small amount of mercury and an inert gas, like argon, kept under very low pressure. In these electric-discharge lamps, a fluorescing coating on the glass, called phosphor powder, transforms some of the ultraviolet energy generated into light. Fluorescent lamps also require a ballast to start and maintain their operation. Early fluorescent lamps were sometimes criticized as not producing enough warm colors, making them appear as too white or uncomplimentary to skin tones, and a cool white, fluorescent lamp had a CRI of 62. But today, there are lamps available with CRIs of 80 and above that simulate natural daylighting and incandescent light. They are also available in a variety of CCTs: 2900 to 7000. The "T" designation for fluorescent lamps stands for tubular—the shape of the lamp. The number after the "T" gives the diameter of the lamp in eighths of an inch. The T8 lamp—available straight or U-shaped—has become the standard for new construction. It also commonly serves as a retrofit replacement for 40-watt T12 lamps, improving efficacy, CRI, and efficiency. In some cases, T10 lamps offer advantages over both T12 and T8 lamps, including higher efficiency, higher CRI values, a wider selection of CCTs, and compatibility with several ballast types. Another lamp type is the T5FT fluorescent lamp. These lamps produce maximum light output at higher ambient temperatures than those that are linear or U-shaped. Linear fluorescent lamps are often less expensive than compact fluorescent lamps. They can also produce great light, are easier to dim, and last longer.

DID YOU KNOW?

Cold cathode fluorescent lamps are one of the latest technological advances in fluorescent technology. The "cold" in cold cathode means there is no heating filament in the lamp to heat up the gas. This makes them more efficient. Also, since there is no filament to break, they are ideal for use in rough service environments where a regular lamp may fail. They are often used as backlights in LCLD monitors. They are also used in exit signs.

Compact Fluorescent Lamps (CFLs)—are small-diameter fluorescent lamps folded for compactness, with an efficacy of 50–75 lm/w for 27–40 watts. There are several styles of CFLs: two-, four-, and six-tube lamps, as well as circular lamps. Some CFLs have the tubes and ballast permanently connected. Others have separate tubes and ballasts. Some CFLs feature a round adapter, allowing them to screw into common electrical sockets and making them ideal replacements for incandescent lamps. They last up to ten times longer than incandescent lamps, and they use about one-fourth the energy, producing 90% less heat. However, typical 60–100-watt incandescent lamps are no more than 5.3 inches long, while standard CFLs are longer than six inches. Therefore, sub-CFLs have been developed. No more than 4.5 inches long, sub-CFLs fit into most incandescent fixtures.

DID YOU KNOW?

Because of their energy efficiency, brightness, and low heat output, CFLs are also good replacements for halogen lamps in torchieres.

High Intensity Discharge Lamps (HID)—compared to fluorescent and incandescent lamps, high-intensity discharge (HID) lamps produce a large quantity of light in a small package. HID lamps produce light by striking an electrical arc across tungsten electrodes housed inside a specially designed inner glass tube. This tube is filled with both gas and metals. The gas aids in the starting of the lamps. Then, the metals produce the light once they are heated to a point of evaporation. Like fluorescent lamps, HID lamps require a ballast to start and maintain their operation. Types of HID lamps include mercury vapor (CRI range 15–55), metal halide (CRI range 65–80), and high-pressure sodium (CRI range 22–75). Mercury vapor lamps (25–45 lm/w), which originally produced a bluish-green light, were the first commercially available HID lamps. Today, they are also available in a color corrected, whiter light. But they are still often being replaced by the newer, more efficient high-pressure sodium and metal halide lamps. Standard high-pressure sodium lamps have the highest efficacy of all HID lamps, but they produce a yellowish light. High-pressure sodium lamps that produce a whiter light are now available, but efficiency is somewhat sacrificed. Metal halide lamps are less efficient but produce an even whiter, more natural light. Colored metal halide lamps are also available.

DID YOU KNOW?

HID lamps are typically used when high levels of light are required over large areas and when energy efficiency and/or long life are desired. These areas include gymnasiums, large public areas, warehouses, outdoor activity areas, roadways, parking lots, and pathways. More recently, however, HID sources—especially metal halide (45–100 lumens/watt)—have been used in small retail and residential environments.

Low Pressure Sodium Lamps—have the highest efficacy of all commercially available lighting sources, producing up to 180 lumens/watt. Even though they emit a yellow light, a low-pressure sodium lamp should not be confused with a standard high-pressure sodium lamp—a high-intensity discharge lamp. Low-pressure sodium lamps operate much like a fluorescent lamp and require a ballast. The lamps are also physically large—about four feet long for the 180-watt size—so light distribution from fixtures is less controllable. There is a brief warm-up period for the lamp to reach full brightness. With a CRI of 0, low-pressure sodium lamps are used where color rendition is not important but energy efficiency is. They are commonly used for outdoor, roadway, parking lot, and pathway lighting. Low-pressure sodium lamps are preferred around astronomical observatories because the yellow light can be filtered out of the random light surrounding the telescope.

Solid-State Lighting (SSL)—compared to incandescent and fluorescent lamps, solid-state lighting creates light with less directed heat. A semiconducting material converts electricity directly into light, which makes the light very energy efficient. Solid-state lighting includes a variety of light producing semiconductor devices including light-emitting diodes (LEDs) and organic light-emitting diodes (OLEDs). Warm white LEDs have an efficacy of 50 lm/w, while cool white LEDs can achieve efficacies up to 100 lm/w. Until recently, LEDs—basically tiny light bulbs that fit easily into an electrical circuit—were used as simple indicator lamps in electronics and toys. But recent research has achieved efficiencies equal to fluorescent lamps. And the cost of semiconductor material, which used to be quite expensive, has lowered, making LEDs a more cost-effective lighting option. Ongoing research shows that LEDs have great potential as energy-efficient lighting for residential and commercial building use. New uses for LEDs include small area lighting, such as task and under-shelf fixtures, decorative lighting, and pathway and step marking. As LEDs become more powerful and effective, they will be used in more general illumination applications, perhaps with entire walls and ceilings becoming the lighting system. They are already being used successfully in many general illumination applications including traffic signals and exit signs. OLEDs currently are used in very thin, flat display screens, such as those in portable televisions, some vehicle dashboard readouts, and in postage-stamp-sized data screens built into pilots' helmet visors. Because OLEDs emit their own light and can be incorporated into arrays on very thin, flexible materials, they also could be used to fashion large, extremely thin panels for light sources in buildings.

DID YOU KNOW?

RBG designation—red, blue, green. One way to create white light with LEDs is to mix the three primary colors of light.

PC designation—phosphor conversions. White light can be produced by a blue, violet, or near-UV LED coated with yellow or multi-chromatic phosphors. The combined light emission appears white.

BALLASTS

Ballasts consume, transform, and control electrical power for electric-discharge lamps, providing the necessary circuit conditions for starting and operating them. Electric-discharge lamps include fluorescent, high-intensity discharge, and low-pressure sodium. When comparing ballasts, it is important to understand the following performance characteristics:

- Ballast Factor (BF)—the ratio of the slight output of a lamp or lamps operated by a specific ballast to the light output of the same lamp(s) operated by a reference ballast. It can be used to calculate the actual light output of that specific lamp-ballast combination. BF is typically different for each lamp type. Ballasts with extremely high BFs could reduce lamp life and accelerate lumen deficiency because of high lamp current. Extremely low BFs also could reduce lamp life because they reduce lamp current.
- Ballast Efficacy Factor (BEF)—the ratio of ballast factor (as a percentage) to power (in watts). BEF comparisons should be made only among ballasts operating the same type and number of lamps.
- System Efficacy—the ratio of the light output to the power, measured in lumens per watt (lm/w), for a particular lamp ballast system.

There are three basic types of ballasts: Magnetic, hybrid, and electronic.

Magnetic Ballasts—contain a magnetic core of several laminated steel plates wrapped with copper windings. These ballasts operate lamps at line frequency (60 hertz in North America). Of all ballasts, magnetic ones are the least expensive and the least efficient. They have greater power losses than electronic ballasts. But magnetic ballasts manufactured today are 10% more efficient than the older high-loss magnetic ballasts, which used aluminum windings. Magnetic ballasts are available with dimming capability. However, they cannot be dimmed below 20% and still use more electricity than electronic ballasts.

Hybrid Ballasts (Cathode-Disconnect Ballasts)—use a magnetic core-and-coil transformer and an electronic switch for the electrode heating circuit. Like magnetic ballasts, they operate lamps at the frequency (60 hertz in North America). After they start the lamp, these ballasts discount the electrode-heating circuit. Hybrid ballasts cost more than magnetic ballasts, but they are more energy efficient.

Electronic Ballasts—in the early 1980s, manufacturers began to replace the core-and-coil transformer with solid-state, electronic compounds that could operate lamps at 20–50 kilohertz. These electronic ballasts experience half the power loss of magnetic ballasts. Also, lamp efficacy increases by approximately 10% to 15% compared to 60-hertz operation. Electronic ballasts are the most expensive, but they are also the most efficient. Operating lamps with electronic ballasts reduces electricity use by 10% to 15% over magnetic ballasts of the same light output. They are also quieter, lighter, and they virtually eliminate lamp flicker. Electronic ballasts are also available as dimming ballasts. These ballasts allow the light level to be controlled between 1% and 100%. There are a variety of electronic ballasts available for use with fluorescent lamps. Electronic ballasts have been successfully used with lower watt high-intensity discharge (HID) lamps (primarily 35–100 w/MH). These ballasts provide an energy savings over magnetic ballasts of 8% to 20%. Their lighter weight also helps in some HID applications, such as track lighting.

LUMINAIRES (LIGHTING FIXTURES)

A lamp or lamp-ballast combination may produce light very efficiently, but if it is installed in an inefficient luminaire, the overall system efficiency may still be poor. The best luminaire manufacturers will design their fixtures around specific lamps to optimize the amount of light delivered to the work area. For example, a luminaire designed specifically for a compact fluorescent lamp can deliver almost ten times as much illumination as an incandescent fixture fitted with the same compact fluorescent lamp.

Luminaire components include reflectors, diffusers—which absorb some of the light from a lamp—and polarizing panels. Reflectors can be used to direct more of the light produced by the lamp out of the luminaire onto the work area. Polarizing panels can sometimes increase the contrast of a visual task.

When comparing luminaires, it is important to understand the flowing performance characteristics:

- Illuminance—the amount of light that reaches a surface. It is measured in footcandles (lumens/square foot) or lux (lumens/square meter).
- Luminaire Efficacy/Efficiency Rating (LER)—the light output (lumens) per watt of electricity use (lm/w).

Before selecting luminaires or lighting fixtures for an office building, factory, warehouse, or even a parking lot, it is a good idea to consult a certified lighting designer. A lighting designer will not only help find the most energy-efficient luminaires, but also provide lighting that makes for a comfortable and more productive work environment.

Today, energy-efficient commercial lighting design includes more than just the ambient or general lighting of a workspace, such as the use of ceiling luminaires. When designing or retrofitting the lighting, the general illuminance can be reduced if task lighting is implemented properly into the overall design. Task lighting can result in significant energy savings and improved visibility for workers.

DID YOU KNOW?

The most efficient light source technology for exit signs is light-emitting diodes (LED). The most popular parking lot luminaires use energy-efficient, high-intensity discharge lamps or low-pressure sodium lamps. But the most efficient light source technology for outdoor use is outdoor photovoltaic lighting.

LIGHTING CONTROLS

Lighting controls help conserve energy and make a lighting system more flexible. The most common light control is the on/off switch. Other types of light control technologies include manual dimming, photosensors, occupancy sensors, clock switches or timers, and centralized controls.

Manual dimming—these controls allow occupants of a space to adjust the light output or illuminance. This can result in energy savings through reductions in input power, as well as reductions in peak power demand, and enhanced lighting flexibility. Slider switches allow the occupant to change the lighting over the complete output range. They are the simplest of the manual controls. Preset scene controls change the dimming setting for various lights all at once with the press of a button. It is possible to have different settings for the morning, afternoon, and evening. Remote control dimming is also available. This type of technology is well suited for retrofit projects, where it is useful to minimize rewiring.

DID YOU KNOW?

Fluorescent lighting fixtures require special dimming ballasts and compatible control devices. Some dimming systems for high-intensity discharge lamps also require special dimming ballasts.

Photosensors—these devices automatically adjust the light output of a lighting system based on detected illuminance. The technology behind photosensors is the photocell. A photocell is a light-responding silicon chip that converts incident radiant energy into electrical current. While some photosensors just turn lights off and on, others can also dim lights. Automatic dimming can help with lumen maintenance. Lumen maintenance involves dimming luminaires when they are new, which minimizes the wasteful effects of overdesign. The power supplied to them is gradually increased to compensate for light loss over the life of the lamp.

DID YOU KNOW?

Nearly all photosensors are used to decrease the electric power for lighting. In addition to lowering the electric power demand, dimming the lights also reduces the heat load on a building's cooling system. Any solar heat gain that occurs in a building during the day must be considered for a whole building energy usage analysis.

Occupancy sensors—turn lights on and off based on their detection of motion within a space. Some sensors can also be used in conjunction with dimming controls to keep the lights from turning completely off when a space is unoccupied. This control scheme may be appropriate when occupancy sensors control separate zones in a large space, such as in a laboratory or in an open office area. In these situations, the lights can be dimmed to a predetermined level when the space is unoccupied. Sensors can also be used to enhance the efficiency of centralized controls by switching off lights in unoccupied areas during normal working hours as well as after hours. There are three basic types of occupancy sensors, these include:

- Passive infrared (PIR)—these sensors react to the movement of a heat-emitting body through their field of view. Wall box-type PIR occupancy sensors are best suited to small, enclosed spaces such as private offices, where the sensor replaces the light switch on the wall and no extra wiring is required. They should not be used where walls, partitions, or other objects might block the sensors' ability to detect motion.
- Ultrasonic—these sensors emit an inaudible sound pattern and reread the reflection. They react to changes in the reflected sound pattern. These sensors detect very minor motion better than most infrared sensors. Therefore, they are good to use in spaces such as restrooms with stalls, which can block the field of view, because the hard surfaces will reflect the sound pattern.
- Dual-technology (hybrid)—these are occupancy sensors that use both passive infrared and ultrasonic technologies to minimize the risk of false triggering (lights coming on when the space is unoccupied). They also tend to be more inexpensive.

Clock switches or timers—These control lighting for a preset period. They come equipped with an internal mechanical or digital clock, which will automatically adjust for the time of year. The user determines when the light should be turned on and when it should be turned off. Clock switches can be used in conjunction with photosensors.

Centralized controls—these are building controls or building automation systems that can be used to automatically turn on, turn off, or dim electrical lights around a building. In the morning, the centralized control system can be used to turn on the lights before employees arrive. During the day, a central control system can be used to dim the lights during periods of high-power demand. And, at the end of the day, the lights can be turned off automatically. A centralized lighting control system can significantly reduce energy use in buildings where lights are left on when not needed.

DAYLIGHTING

As mentioned earlier, daylighting is the practice of placing windows or other openings and reflective surfaces so that during the day, natural light provides effective internal lighting. When properly designed and effectively integrated with the electric lighting system, daylighting can offer significant energy savings by offsetting a portion of the electric lighting load. A related benefit is the reduction in cooling capacity and use by lowering a significant component of internal gains. In addition to energy savings, daylighting generally improves occupant satisfaction and comfort. Windows also provide visual relief, contact with nature, time orientation, the possibility of ventilation, and emergency egress.

The Daylight Zone—high daylight potential is found particularly in those spaces that are predominantly daytime occupied. Site solar analysis should assess the access to daylight by considering what is "seen" from the various potential window orientations. What proportion of the sky is seen from typical task locations in the room? What are the exterior obstructions and glare sources? Is the building design going to shade a neighboring building or landscape feature that is dependent on daylight or solar access? It is important to establish which spaces will most benefit from daylight and which spaces have little or no need for daylight. Within the spaces that can use daylight, place the most critical visual tasks in positions near the window. Try to group tasks by similar lighting requirements and occupancy patterns. Avoid placing the windows in the direct line of sight of the occupant as this can cause extreme contrast and glare. It is best to orient the occupant at 90 degrees from the window. Where privacy is not a major concern, consider interior glazing (known as relights or borrowed lights) that allow light from one space to be shared with another. This can be achieved with transom lights, vision glass, or translucent panels if privacy is required. The floor plan configuration should maximize the perimeter daylight zone. This may result in a building with a higher skin-to-volume ratio than a typical compact building design. A standard widow can produce useful illumination to a depth of about 1.5 times the height of the window. With light shelves or other reflector systems this can be increased to 2.0 times or more. As a general rule of thumb, the higher the window is placed on the wall, the deeper the daylight penetration.

Window Design Considerations—the daylight that arrives at a work surface comes from three sources:

1. The exterior reflected component. This includes ground surfaces, pavement, adjacent buildings, wide windowsills, and objects. Remember that excessive rough reflectance will result in glare.
2. The direct sun/sky component. Typically, the direct sun component is blocked from occupied spaces because of heat gain, glare, and UV degradation issues. The sky dome then becomes an important contribution to daylighting the space.

3. The internal reflected component. Once the daylight enters the room, the surrounding walls, ceiling, and floor surfaces are important light reflectors. Using high-reflectance surfaces will better bounce the daylight around the room and it will reduce extreme brightness contrast. Window frame material should be light-colored to reduce contrast with the view and have a non-specular finish to eliminate glare spots. The window jambs and sills can be beneficial light reflectors. Deep jambs should be splayed (angled toward the interior) to reduce the contrast around the perimeter of the window.

DID YOU KNOW?

The most important interior light-reflecting surface is the ceiling. High-reflectance paints and ceiling tiles are now available with .90 or higher reflectance values. Tilting the ceiling plant toward the daylight source increases the daylight that is reflected from this surface. In small rooms there should also be a high reflectance matte finish. The sidewalls, followed by the floor have less impact on the reflected daylight in the space.

Major room furnishings such as office cubicles or partitions can have a significant impact on reflected light, so select light-colored materials. Suggested reflectance levels for various room surfaces are:

*   Ceilings: > 80%
*   Walls: 50%–70%
*   Floors: 20%–40%
*   Furnishings: 25%–45%

Because light essentially has no scale for architectural purposes, the proportions of the room are more important than the dimensions. A room that has a higher ceiling compared to the room depth will have deeper penetration of daylight whether from side lighting (windows) or top lighting (skylights and clerestories). Raiding the window head height will also result in deeper penetration and more even illumination in the room. Punched window openings, such as small, square windows separated by wall area, result in uneven illumination and harsh contrast between the window and adjacent wall surfaces. A more even distribution is achieved with horizontal strip windows.

DID YOU KNOW?

There is no direct sunlight on the polar-side wall of a structure from the autumnal equinox to the spring equinox in parts of the globe north of the Tropic of Cancer and in parts of the globe south of the Tropic of Capricorn.

Effective Aperture (EA)—One method of assessing the relationship between visible light and the size of the window is the effective aperture method. The *effective aperture* (EA) is defined as the product of the visible transmittance and the window-to-wall ratio (WWR). The window-to-wall ratio is the proportion of window area compared to the total wall area where the window is located. For example, if a window covers 25 square feet in a 100-square-foot wall, then the WWR is 25/100, or 0.25. A good starting target for EA is in the range of 0.20 to 0.30. For a given EA number, a higher WWR (larger window) results in a lower visible transmittance.

Example: WWR = .5 (half the wall in glazing)

VT = .6, EA = 0.3

Or WWR = .75, VT = .4 for the same EA of 0.3

DID YOU KNOW?

Typically, lowering the visible transmittances will also lower the shading coefficient, but you must verify this with glazing manufacturer data since this is not always the case.

Light Shelves—are effective horizontal light-reflecting overhangs placed above eye level with a transom window placed above them. Light shelves enhance the lighting from windows of the equator-facing side of a building. Exterior shelves are more effective shading devices than interior shelves. A combination of exterior and interior will work best in providing an even illumination gradient.

Top Lighting Strategies—large single-level floor areas and the top floors of multistory buildings can benefit from top lighting. The general types of top lighting include skylights, clerestories, monitors, and sawtooth roofs.

- Skylights—horizontal skylights can be an energy problem because they tend to receive maximum solar gain at the peak of the day. The daylight contribution also peaks at midday and falls off severely in the morning and afternoon. There are high-performance skylight deigns that incorporate reflectors or prismatic lenses that reduce the peak daylight and heat gain while increasing early and late afternoon daylight contributions. Another option is light pipes where a high reflectance duct channels the light from a skylight down to a diffusing lens in the room. These may be advantageous in deep roof constructions.
- Clerestory window—is a vertical glazing located high on an interior wall. South-facing clerestories can be effectively shaded from direct sunlight by a properly designed horizontal overhang. In this design the interior north wall can be sloped to better reflect the light down into the room. Use light-colored overhangs and adjacent roof surfaces to improve the reflected component. If exterior shading is not possible, consider interior vertical baffles to better diffuse the light. A south-facing clerestory will produce higher daylight illumination than a north-facing clerestory. East- and west-facing clerestories have

the same problems as east and west windows: difficult shading and potentially high heat gains.

- Roof monitor—consists of a flat roof section raised above the adjacent roof with vertical glazing on all sides. This design often results in excessive glazing area, which results in higher heat losses and gains than a clerestory design. The multiple orientations of the glazing can also create shading problems
- Sawtooth roof—is an old design often seen in industrial buildings. Typically, one sloped surface is opaque and the other is glazed. A contemporary saw-tooth roof may have solar collectors or photovoltaic cells on the south-facing slope and daylight glazing on the north-facing slope. Unprotected glazing on the south-facing sawtooth surface may result in high heat gains. In these applications an insulated diffusing panel may be a good choice.

DID YOU KNOW?

A building designed for daylighting but without an integrated electrical lighting system will be a net energy loser because of the increased thermal loads. Only when the electric lighting load is reduced will there be more than offsetting savings in electrical and cooling loads. The benefits from daylighting are maximized when both occupancy and lighting sensors are used to control the electric lighting system.

THE BOTTOM LINE ON SOLAR ENERGY

Solar energy has some obvious advantages in that the source is free; however, the initial investment in operating equipment is not free or inexpensive. Solar energy is also environmentally friendly, requires almost no maintenance, and reduces our dependence on foreign energy supplies. Probably the greatest downside of solar energy use is that in areas without direct sunlight during certain times of the year, solar panels cannot capture enough energy to provide heat for home or office. Geographically speaking, the higher latitudes do not receive as much direct sunlight as tropical areas. Because of the position of the sun in the sky, solar panels must be placed in sun-friendly locations such as the US Desert Southwest and the Sahara region of northern Africa. Another downside of solar energy is the efficiency of the system; it may be seriously affected by how well it was installed.

The real bottom line of solar energy is obvious; its advantages outnumber its disadvantages.

NOTES

1 Source: Galileo Quote (2023). Accessed 8/10/23 @ www.thoughttco.com/galileo-galilei-quotes.
2 Source: Biography/Personal Quotes. Accessed 6/24/23 @ www.imdb.com.

3 Most of the information in this section is from EIA's (2022) *Concentrating Solar Power: Energy from Mirrors.* Washington, DC: US Department of Energy.
4 Case study is found in the USDOE's Federal Technology Alert on Transpired Collectors. Accessed 02/10/23 @ www.eren.doe.gov/solarbuildings.
5 Much of the information in this section is from USDOE: EERE (2008) *Energy Efficiency and Renewable Energy.* Accessed 02/13/20 @ www1.eere.energy.gov/buildings/commerc ial/printable_versions/lighting.html.

REFERENCES AND RECOMMENDED READING

Chiras, D.D. 2002. *The Solar House: Passive Heating and Cooling.* White River Junction, VT: Chelsea Green, p. 33.

Duffie, J. and Beckman, W. 1991. *Solar Engineering of Thermal Processes.* New York: John Wiley & Sons.

Energy Efficiency & Renewable Energy (EERE). 2008a. *Solar Energy Technologies Program.* Accessed 02/04/10 @ www.eere.energy.gov

Energy Efficiency & Renewable Energy (EERE). 2008b. *Linear Concentrator Systems.* Accessed 02/04/21 @ www.EERE.energy.gov/solar/printable_versions/linear_concen trators.html

Energy Efficiency & Renewable Energy (EERE). 2008c. *Thermal Storage.* Accessed 02/08/19 @ www.eere.energy.gov/solar/ printable_versions/thermal_storage.html

Enermodal Engineering Limited. 1995. *Performance of the Perforated-Plate/Canopy Solarwall at GM Canada, Oshawa.* Ottawa, ON: Energy Technology Branch, Department of Natural Resources.

Hanson, B.J., 2004. *Energy Power Shift: Benefiting from Today's New Technologies.* Maple, WI: Lakota Scientific Press.

Hinrichs and and Kleinbach. 2006. *Energy: Its Use and the Environment.* Boston, MA: Brooks/ Cole.

National Renewable Energy Laboratory (NREL). 2009. *Concentrating Solar Power.* Accessed 06/13/19 @ www.nrel.gov/learning/re_csp.html

National Renewable Energy Laboratory (NREL). 2010. *CSP Systems.* Accessed 02/07/10 @ www.nrel.gov/concentratingsolar.html

Patel, M. 1999. *Wind and Solar Power Systems.* Boca Raton, FL: CRC Press.

Perez-Blanco, H. 2009. *The Dynamics of Energy: Supply, Conversion, and Utilization.* Boca Raton, FL: CRC Press.

US Department of Energy (USDOE). 1998. *Solar Dish/Engine Systems.* Washington, DC: US Department of Energy.

US Department of Energy (USDOE). 2006. *Solar Buildings: Transpired Air Collectors.* Accessed 02/11/10 @ www.eren.doe.Gov/solarbuildings

US Department of Energy (USDOE). 2008. *Concentration Solar Power Commercial Application Study.* Washington, DC: US Dept. of Energy.

US Department of Energy (USDOE). 2009a. *Solar Energy Technologies Program.* Accessed 06/13/19 @ www1.eere.Energy.gov/solar/printable_versions/about.html

US Department of Energy (USDOE). 2009b. *Fossil Fuels.* US Dept. of Energy. Accessed 06/ 10/09 @ www.energy.gov/energysources/fossilfuels.htm

3 Wind Energy

INTRODUCTION[1]

The Good, the Bad, and the Ugly of Wind Energy:

Good: As long as Earth exists, the wind will always exist. The energy in the winds that blow across the United States each year could produce more than 16 billion GJ of electricity—more than one and one-half times the electricity consumed in the United States in 2000.

Bad: Turbines are expensive. Wind doesn't blow all the time, so they must be part of a larger plan. Turbines make noise. Turbine blades kill birds.

Ugly: Some look upon giant wind turbine blades cutting through the air as grotesque scars on the landscape, as visible polluters.

The bottom line: Do not expect Don Quixote, mounted in armor on his old nag, Rocinante, with or without Sancho Panza, to charge those windmills. Instead, expect—you can count on it, bet on it, and rely on it—that the charge to build those windmills will be done by the rest of us, to satisfy our growing, inexorable need for renewable energy. What other choice do we have?

Obviously, wind energy or power is all about wind. In simple terms, wind is the response of the atmosphere to uneven heating conditions. Regarding Earth's atmosphere, and, again, to state the obvious, it is constantly in motion. Anyone observing the constant weather changes and cloud movement around them is aware of this phenomenon. Although its physical manifestations are obvious, the importance of the dynamic state of our atmosphere is much less obvious.

The constant motion of Earth's atmosphere (air movement) consists of both horizontal (*wind*) and vertical (*air currents*) dimensions. The atmosphere's motion is the result of thermal energy produced from the heating of Earth's surface and the air molecules above. Because of differential heating of Earth's surface, energy flows from the equator poleward.

Hanson (2004) points out that the energy resources contained in the wind in the United States are well known and mapped in detail. Even though this is the case and air movement plays a critical role in transporting the energy of the lower atmosphere, bringing the warming influences of spring and summer and the cold chill

DOI: 10.1201/9781003439059-4

43

of winter, and generally that wind and air currents are fundamental to how nature functions, the effects of air movements on our environment are often overlooked. All life on Earth has evolved or has been sustained with mechanisms dependent on air movement: pollen is carried by winds for plant reproduction; animals sniff the wind for essential information; wind power was the motive force that began the earliest stages of the Industrial Revolution. Now we see the effects of winds in other ways, too: Wind causes weathering (erosion) of the Earth's surface; wind influences ocean currents; air pollutants and contaminants such as radioactive particles transported by the wind impact our environment.

DID YOU KNOW?

Wind speed is generally measured in m/s, but Americans usually think in mph. A good rule of thumb to keep at hand is that to convert m/s to mph, double the value in m/s and add 10%.

AIR IN MOTION

In all dynamic situations, forces are necessary to produce motion and changes in motion—winds and air currents. The air (made up of various gases) of the atmosphere is subject to two primary forces: gravity and pressure differences from temperature variations.

Gravity (gravitational forces) holds the atmosphere close to the Earth's surface. Newton's law of universal gravitation states that everybody in the universe attracts another body with a force equal to:

$$F = G\frac{m_1 m_2}{r2},$$
(3.1)

where:
F = Force,
m_1 and m_2 = the masses of the two bodies,
G = universal constant of 6.67×10^{-11} N x m^2/kg^2, and
R = distance between the two bodies.

The force of gravity decreases as an inverse square of the distance between them. Thermal conditions affect density, which in turn causes gravity to affect vertical air motion and planetary air circulation. This affects how air pollution is naturally removed from the atmosphere.

Although forces in other directions often overrule gravitational force, the ever-present force of gravity is vertically downward and acts on each gas molecule, accounting for the greater density of air near the Earth.

Atmospheric air is a mixture of gases, so the gas laws and other physical principles govern its behavior. The pressure of a gas is directly proportional to its temperature. Pressure is force per unit area (P = F/A), so a temperature variation in air generally gives rise to a difference in pressure of force. This difference in pressure resulting from temperature differences in the atmosphere creates air movement—on both large and local scales. This pressure difference corresponds to an unbalanced force, and when a pressure difference occurs, the air moves from a high- to a low-pressure region.

In other words, horizontal air movements (called *advective winds*) result from temperature gradients, which give rise to density gradients and subsequently, pressure gradients. The force associated with these pressure variations (*pressure gradient force*) is directed at right angles to (perpendicular to) lines of equal pressure (called *isobars*) and is directed from high to low pressure.

In Figure 3.1, the pressures over a region are mapped by taking barometric readings at different locations. Lines drawn through the points (locations) of equal pressure are called isobars. All points on an isobar are of equal pressure, which means no air movement along the isobar. The wind direction is at right angles to the isobar in the

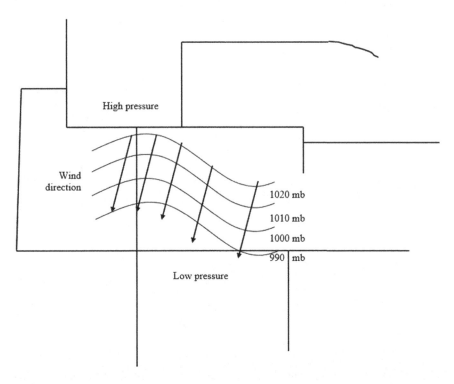

FIGURE 3.1 Isobars drawn through locations having equal atmospheric pressures. The air motion, or wind direction, is at right angles to the isobars and moves from a region of high pressure to a region of low pressure.

direction of the lower pressure. In Figure 3.1, notice that air moves down a pressure gradient toward a lower isobar like a ball rolls down a hill. If the isobars are close together, the pressure gradient force is large, and such areas are characterized by high wind speeds. If isobars are widely spaced (see Figure 3.1), the winds are light because the pressure gradient is small.

Localized air circulation gives rise to *thermal circulation* (a result of the relationship based on a law of physics whereby the pressure and volume of a gas is directly related to its temperature). A change in temperature causes a change in the pressure and/or volume of a gas. With a change in volume comes a change in density, since $P = m/V$, so regions of the atmosphere with different temperatures may have different air pressures and densities. As a result, localized heating sets up air motion and gives rise to *thermal circulation*. To gain understanding of this phenomenon, consider Figure 3.2.

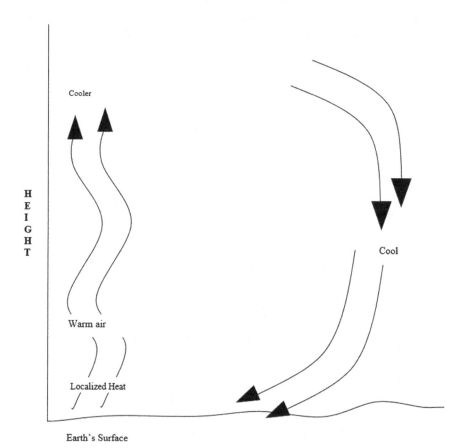

FIGURE 3.2 Thermal circulation of air. Localized heating, which causes air in the region to rise, initiates the circulation. As the warm air rises and cools, cool air near the surface moves horizontally into the region vacated by the rising air. The upper, still cooler, air then descends to occupy the region vacated by the cool air.

Once the air has been set into motion, secondary forces (velocity-dependent forces) act. These secondary forces are caused by Earth's rotation (*Coriolis force*) and contact with the rotating Earth (friction). The Coriolis force, named after its discoverer, French mathematician Gaspard Coriolis (1772–1843), is the effect of rotation on the atmosphere and on all objects on the Earth's surface. In the Northern Hemisphere, it causes moving objects and currents to be deflected to the right; in the Southern Hemisphere, it causes deflection to the left, because of the Earth's rotation. Air, in large-scale north or south movements, appears to be deflected from its expected path. That is, air moving poleward in the Northern Hemisphere appears to be deflected toward the east; air moving southward appears to be deflected toward the west.

The Coriolis effect on a propelled particle (analogous to the apparent effect of an air mass flowing from Point A to Point B). The action of the Earth's rotation on the air particle as it travels north over the Earth's surface as Earth rotates beneath it from east to west—because the Earth rotates east to west beneath it.

Friction (drag) can also cause the deflection of air movements. This friction (resistance) is both internal and external. The friction of its molecules generates internal friction. Friction is also generated when air molecules run into each other. External friction is caused by contact with terrestrial surfaces. The magnitude of the frictional force along a surface is dependent on the air's magnitude and speed, and the opposing frictional force is in the opposite direction of the air motion.

WIND ENERGY[2]

Wind energy is the movement of wind to create power. Since early recorded history, people have been harnessing the energy of the wind for milling grain, pumping water, and for other mechanical power applications. Wind energy propelled boats along the Nile River as early as 5000 BCE. By 200 BCE, simple windmills in China were pumping water, while vertical-axis windmills with woven reed sails were grinding grain in Persia and the Middle East.

The use of wind energy spread around the world and by the eleventh century, people in the Middle East were using windmills extensively for food production; returning merchants and crusaders carried this idea back to Europe. The Dutch refined the windmill and adapted it for draining lakes and marshes in the Rhine River Delta. When settlers took this technology to the Americas in the late nineteenth century, they began using windmills to pump water for farms and ranches, and later, to generate electricity for homes and industry. The first known wind turbine for producing electricity was by Charles F. Brush, in Cleveland, Ohio, in 1888; it was a 12-kW unit used to charge batteries in the cellar of a mansion. The first wind turbine outside of the United States to generate electric was built by Poul la Cour in Denmark in 1891. Used electricity from wind turbines is used to electrolyze water to make hydrogen for the gas lights at the school house. By the 1930s and 1940s in the United States, hundreds of thousands of wind turbines were in use in rural areas not yet served by the grid. The oil crisis in the 1970s created a renewed interest in wind until the US government stopped giving tax credits. Today, there are several hundred thousand windmills in operation around the world, many of which are used for water pumping. But it is the use of wind energy as a pollution-free means of generating electricity on a significant scale that is attracting most current interest in the subject. As a matter of fact, with the present and pending shortage and high cost of fossil fuels to generate electricity and the green movement toward the use of cleaner fuels, wind energy is the world's fastest-growing energy source and will power industry, businesses, and homes with clean, renewable electricity for many years to come. In the United States since 1970, wind-based electricity generating capacity has increased markedly, although (at present) it remains a small fraction of total electric capacity and consumption (see Table 3.1). But this trend is beginning to change—with the advent of $4/gal of gasoline, high heating and cooling costs and subsequent increases in the cost of electricity, worldwide political unrest, and uncertainty in oil-supplying countries, one only needs to travel the "wind corridors" of the United States encompassing parts of Arizona, New Mexico, Texas, Missouri, and north through the Great Plains to the Pembina Escarpment and Turtle Mountains of North Dakota and elsewhere, to witness the considerable activity the seemingly exponential increase in wind energy development and wind turbine installations; these machines are being installed to produce and provide electricity to the grid.

Table 3.1 highlights wind energy's quadrillion Btu ranking in current renewable energy source use. As pointed out in the table, the energy consumption by energy source consumption for the year 2022 has increased. Thus, the 27% wind power consumption figure is expected to steadily increase to an increasingly higher level and this should be reflected in the 2022–2023 figures when they are released.

TABLE 3.1
US Energy Consumption by Energy Source, 2022 (%)

Energy Source	2022
Total	**97.33**
Renewable	12.16
Biomass (biofuels, waste, wood and wood-derived)	40%
Biofuels	19%
Waste	4%
Wood-derived Fuels	17%
Geothermal	2%
Hydroelectric Conventional	19%
Solar/PV	12%
Wind	**27%**

Source: EIA 2022. *US Energy consumption by Energy Source.* Accessed 06/25/23 @ www.eia.energyexplained/us-energy-facts.

DID YOU KNOW?

We can classify wind energy as a form of solar energy. As mentioned, winds are caused by uneven heating of the atmosphere by the sun, irregularities of the Earth's surface, and the rotation of the Earth. As a result, winds are strongly influenced and modified by local terrain, bodies of water, weather patterns, vegetative cover, and other factors. The wind flow, or motion of energy when harvested by wind turbines, can be used to generate electricity.

WIND POWER BASICS

The term "wind energy" or "wind power" describes the process by which the wind is used to generate mechanical power or electricity. Wind turbines convert the kinetic energy in the wind into mechanical power. This mechanical power can be used for specific tasks (such as grinding grain or pumping water), or a generator can convert this mechanical power into electricity (EERE, 2006a).

As mentioned, we have been harnessing the wind's energy for hundreds of years. From old Holland to farms in the United States, windmills have been used for pumping water or grinding grain. Today, the windmill's modern equivalent—a wind turbine—can use the wind's energy to generate electricity. These machines accomplish this like aircraft propeller blades turning in the moving air and powering an electrical generator that supplies an electrical current. Unlike fans that use electricity to make wind, wind turbines use wind to make electricity. The wind turns the blades, which spin a shaft, that connects to a generator and makes electricity (WINDEIS, 2009).

DID YOU KNOW?

Whenever wind energy is being considered as a possible source of renewable energy, it is important to consider the amount of land area required, accessibility to generators, and aesthetics.

WIND TURBINE TYPES

Whether called wind-driven generators, wind generators, wind turbines, wind-turbine generator (WTG), or wind energy conversion system (WECS), modern wind turbines fall into two basic groups; the horizontal-axis (HAWT) variety, like the traditional farm windmills used for water pumping (see Figure 3.3), and the vertical-axis (VAWT) design, like the eggbeater-style Darrieus rotor model, named after its French inventor, the only vertical axis machine with any commercial success. Wind hitting the vertical blades, called aerofoils, generates lift to create rotation. No yaw (rotation about vertical axis) control is needed to keep them facing into the wind. The heavy machinery in the nacelle (cowling) is located on the ground. Blades are closer to ground where

FIGURE 3.3   Typical farm windmill used for pumping water.

wind speeds are lower. Most large modern wind turbines are horizontal-axis turbines; therefore, they are highlighted and described in detail in this text.

DID YOU KNOW?

Wind turbines are grouped in what is called a "wind farm" or a "wind park."

HORIZONTAL-AXIS WIND TURBINES (HAWT)

Wind turbines are available in a variety of sizes, and therefore power ratings. Utility-scale turbines range in size from 100 kilowatts to as large as several megawatts. Horizontal-axis wind turbines (HAWTs) typically either have two or three blades. "Downwind" HAWTs have a turbine with the blades behind (downwind from) the tower. No yaw control is needed because they naturally orient themselves in line with the wind. However, these downwind HAWTs undergo a shadowing effect in that when a blade swings behind the tower, the wind it encounters is briefly reduced and the blade flexes. "Upwind" HAWTs usually have three-blades in front of (upwind of) the tower. These upwind wind turbines require a somewhat complex yaw control to keep them facing into the wind. They operate more smoothly and deliver more power and thus are the most used modern wind turbines. The largest machine has blades that span more than the length of a football field, stands 20 building stories high, and produces enough electricity to power 1,400 homes.

INSIDE THE HAWT

Basically, a HAWT consists of three main parts: a turbine, a nacelle, and a tower. Several other important parts are contained within the tower and the nacelle, including the anemometer, blades, brake, controller, bear box, generator, high-speed shaft, low-speed shaft, pitch, rotor, tower, wind direction, wind vane, yaw drive and yaw motor (see Figure 3.4; Alternative Energy News, 2010).

TURBINE COMPONENTS/PARAMETERS

- *Anemometer*—measures the wind speed and transmits wind speed data to the controller.
- *Blades*—Most turbines have either two or three blades. Wind blowing over the blades causes the blades to "lift" and rotate; they capture the kinetic energy of the wind and help the turbine rotate.
- *Brake*—a disc brake, which can be applied mechanically, electrically, or hydraulically to stop the rotor in emergencies.

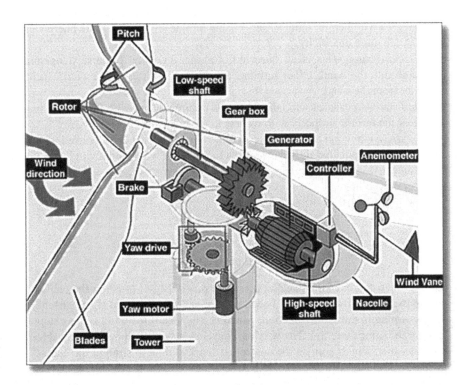

FIGURE 3.4   Horizontal-axis wind turbine components.

- *Controller*—starts up the machine at wind speeds of about 8 to 16 miles per hour (mph) and shuts off the machine at about 55 mph. Turbines do not operate at wind speeds above approximately 55 mph because they might be damaged by the high winds.
- *Gear Box*—gears connect the low-speed shaft to the high-speed shaft and increase the rotational speeds from about 30 to 60 rotations per minute (rpm) to about 1000 to 1800 rpm, the rotational speed required by most generators to produce electricity. The gear box is a costly (and heavy) part of the wind turbine and engineers are exploring "direct-drive" generators that operate at lower rotational speeds and don't need gear boxes.
- *Generator*—usually an off-the-shelf induction generator that produces 60-HertzAC electricity.
- *High-speed shaft*—drives the generator.
- *Low-speed shaft*—rotor turns the low-speed shaft at about 30 to 60 rotations per minute.
- *Nacelle*—sits atop the tower and contains the gearbox, generator, low- and high-speed shafts, controller and brake.
- *Pitch*—bales are turned, or pitched, out of the wind to control the rotor speed and keep the rotor turning in winds that are too high or too low to produce electricity.
- *Rotor*—the blades and the hub together are called the rotor.
- *Tower*—made from tubular steel, concrete, or steel lattice. Because wind speed increases with height, taller towers enable turbines to capture more energy and generate more electricity.
- *Wind direction*—this is an "upwind" turbine, so-called because it operates facing into the wind. Other turbines are designed to run "downwind," facing away from the wind.
- *Wind vane*—measures wind direction and communicates with the yaw drive to orient the turbine properly with respect to the wind.
- *Yaw drive*—upwind turbines face into the wind; the yaw drive is used to keep the entire nacelle and thus the rotor facing into the wind as the wind direction changes. Downwind turbines don't require a yaw drive; the wind blows the rotor downwind.
- *Yaw motor*—powers the yaw drive.

DID YOU KNOW?

During the rotation of the nacelle, there is a possibility of twisting the cables inside the tower. The cables all become more and more twisted if the turbine keeps turning in the same direction, which can happen if the wind keeps changing in the same direction. The wind turbine is therefore equipped with a cable twist counter, which notifies the controller that it is time to straight the cables (Khaligh & Onar, 2010)

WIND ENERGY/POWER CALCULATIONS

A wind turbine is a machine that converts the kinetic energy in wind into the mechanical energy of a shaft. Calculating the energy and power avail in the wind relies on knowledge of basics physics and geometry. The kinetic energy of an object is the extra energy that it possesses because of its motion. It is defined as the work needed to accelerate a body of a given mass from rest to its current velocity. Once in motion, a body maintains its kinetic energy unless its speed changes. The kinetic energy of a body is given by the Equation (3.2)

$$KE = 1/2mv^2, \tag{3.2}$$

where
    KE = Kinetic Energy,
    m = mass, and
    v = velocity.

EXAMPLE 3.1 DETERMINING POWER IN THE WIND

Step 1:

To find the kinetic energy of moving air (wind), let's say we have a large packet of wind [i.e., a geometrical package of air passing through the plain of a wind turbine's blades (which sweep out a cross-sectional area A), with a thickness (D); see Figure 3.5] passing through the plane over a given time.

Step 2:

To determine power in wind, we begin by considering the kinetic energy of the packet of air shown in Figure 3.6 along with mass m moving at velocity v, as shown in Equation 3.3. Next, we need to divide by time to get power:

$$\text{Power Though Area A} \frac{1}{2}\left(\frac{m \; passing \; through \; A}{t}\right)v^2. \tag{3.3}$$

Step 3:

The mass flow rate is (ρ is air density—the mass per unit volume of Earth's atmosphere).

$$m = \frac{m \, passing \, A}{t}\rho Av. \tag{3.4}$$

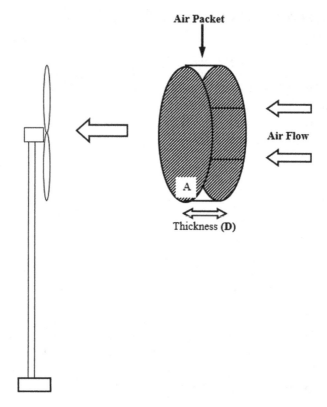

FIGURE 3.5 A packet of air passing through the plane of a wind turbine's blades, with the thickness (D) passing through the plane over a given time.

Step 4:

To obtain the power in the wind equation,

$$\text{Power through area A} = \tfrac{1}{2}\,(\rho A v)v^2$$

$$\text{Power in the wind} = P_w = \tfrac{1}{2}\,\rho A v^3, \qquad (3.5)$$

where

P_w (Watts) = power in the wind,

ρ (kg/m^3) = air density (2.225 kg/m^3 at 15°C and 1 atm),

A (m^2) = cross-sectional area that wind passes through perpendicular to the wind, and

v (m/s) = Windspeed normal to A (1 m/s = 2.237 mph).

Power increases like the cube of wind speed. That is, doubling the wind speed increases the power by eight. Moreover, the energy in 1 hour of 20 mph winds is the same as energy in 8 hours of 10 mph winds. When we speak of power (watts/m^2) and windspeed (mph), we cannot use average wind speed because the relationship is nonlinear. Power in the wind is also proportional to A. For a conventional horizontal-axis wind turbine, A = (π/4)D^2, so wind power is proportional to the blade diameter squared. Because cost is approximately proportional to blade diameter, larger wind turbines are more cost effective.

AIR DENSITY CORRECTION FACTORS

Earlier we pointed out that air density is affected by different temperature and pressure.

Air density correction factors can correct air density for temperature and altitude. Correction factors for both temperature and altitude correction can be found in standardized tables.

Equation (3.6) is used to determine air density for different temperatures and pressures:

$$\rho = \frac{P \times MW \times 10^{-3}}{RT}, \tag{3.6}$$

where
P = absolute pressure (atm) starts at zero gas pressure (no molecules—a vacuum) and has no practical maximum limit,
MW = molecular weight of air (g/mol) = 28.97 g/mol,
T = absolute temperature (K) [The temperature in Kelvin (273 + °C) or Rankine (absolute zero = −460°F, or 0°R)], and
R = ideal gas constant = 8.2056 x 10^{-5} x m^3 x atm x K^{-1} x mol^{-1}.

ELEVATION AND EARTH'S ROUGHNESS

Windspeed is affected by its elevation above the Earth and the roughness of Earth. Because power increases like the cube of windspeed, we can expect a significant economic impact from even a moderate increase in windspeed; thus, in the operation of wind turbines, windspeed is a very important parameter. Earth's surface features can't be ignored when deciding where to place wind turbines and in the calculation of their output productivity. Natural obstructions such as mountains and forests and human-made obstructions such as buildings provide friction as winds flow over them. The point is there is a lot of friction in the first few hundred meters above ground—smooth surfaces (like water) are better. Thus, the greater the elevation above the Earth, the greater the windspeed; tall wind turbine towers are better.

When actual measurements are not possible or available, it is possible to characterize or approximate the impact of rough surfaces and height on wind speed. This is accomplished using Equation (3.7):

$$\frac{v}{v_0} = \left(\frac{H}{H_0}\right)^{\alpha},$$

(3.7)

where
α = friction coefficient—obtained from standardized tables; typical value of α in open terrain is 1/7; for a large city $\alpha = 0.4$; for calm water, $\alpha = 0.1$,
v = windspeed at height H, and
v_0 = windspeed at height H_0 (H_0 is usually 10 m).

DID YOU KNOW?

As pointed out, the energy in wind is a cubed function of wind speed, which means that if the wind speed doubles, there is eight times as much available energy, not twice as much as one might expect.

WIND TURBINE ROTOR EFFICIENCY

Generally, when we think or talk about efficiency we think about input versus output and know that if we put 100% into something and get 100% output, we have a very efficient machine, operation, or process. In engineering, we can approximate efficiency input versus output by performing mass balance calculations. In using these calculations, we know that according to the laws of conservation, materials cannot disappear or be destroyed; thus, they must exist and be somewhere. However, in this input versus output concept, there are a couple of views on maximum rotor efficiency that the ill-informed or misinformed have taken, and neither makes sense but they are stated here to point out what really does make sense. The two wrong assumptions are the following: First, it can be assumed that downwind velocity is zero—the turbine extracted all the power from the wind; second, downwind velocity is the same as the upwind velocity—turbine extracted no power. Again, in a word: Wrong! It was Albert Betz in 1919 who set the record straight when he theorized that there must be some ideal slowing of the wind so that the turbine extracts the maximum power. Betz theorized in his wind turbine efficiency theory (i.e., to their power input vs. power output and overall efficiency in general) that the efficiency rule or laws of conservation, as stated above, do not apply. Betz law states the maximum possible energy that can be derived from a wind turbine. To understand Betz' law and its derivation, we have provided the following explanation.

DERIVATION OF BETZ' LAW

To understand Betz' law, we must first understand the constraint on the ability of a wind to convert kinetic energy in the wind into mechanical power. Visualize wind passing through a turbine (see Figure 3.6)—it slows down and the pressure is reduced, so it expands.

Equation (3.8) is used to determine power extracted by the blades, another step in the derivation of Betz' law:

$$P_b = \tfrac{1}{2}\, m\, (v^2 - v_d^2), \tag{3.8}$$

where
    m = mass flow rate of air within stream tube,
    v = upwind undisturbed windspeed, and
    v_d = downwind windspeed.

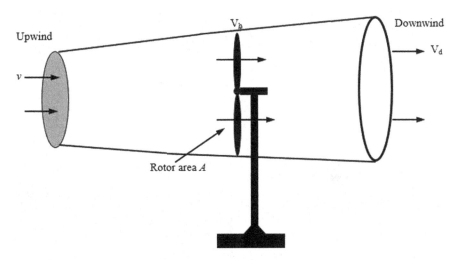

FIGURE 3.6 Wind passing through a turbine.

Using Equation (3.9), determining mass flow rate, is the next step in the Betz' law derivation. In making the determination, it is easiest to use the cross-sectional area A at the plane of the rotor because we know what this value is. The mass flow rate is

$$m = p\, Av_b. \tag{3.9}$$

Assume the velocity through the rotor v_b is the average of upwind velocity v and downwind velocity v_d:

$$v_b = \frac{v + v_d}{2} \quad \longrightarrow \quad m = \rho A\left(\frac{v + v_d}{2}\right)$$

becomes

$$P_b = \frac{1}{2}\rho A\left(\frac{v + v_d}{2}\right)\left(v^2 - v_d^2\right). \tag{3.10}$$

Before moving on in the derivation process, it is important that we define speed/windspeed ratio, λ:

$$\lambda = \frac{v_d}{v}. \tag{3.11}$$

Then we rewrite:

$$P_b = \frac{1}{2}\rho A\left(\frac{v + \lambda v}{2}\right)\left(v^2 - \lambda^2 v^2\right). \tag{3.12}$$

$$\left(\frac{v + \lambda v}{2}\right)\left(v^2 - \lambda^2 v^2\right) = \frac{v^3}{2} - \frac{\lambda^2 v^3}{2} + \frac{\lambda v^3}{2} - \frac{\lambda v^3}{2}$$

$$= \frac{v^3}{2}\left[(1 + \lambda) - \lambda^2(1 + \lambda)\right]$$

$$= \frac{v^3}{2}\left[(1 + \lambda)(1 - \lambda^2)\right]$$

$$\longrightarrow \quad P_b = (\tfrac{1}{2}\, \rho\, Av^3)\, \tfrac{1}{2}\, [(1 + \lambda)(1 - \lambda^2)]$$

$$\underbrace{\hspace{3cm}}_{\downarrow} \quad \underbrace{\hspace{4cm}}_{\downarrow}$$

P_W = Power in the Wind C_P = Rotor Efficiency

The next step is to find the speed windspeed ratio λ, which maximizes the rotor efficiency, C_P:

$$C_P = \frac{1}{2}\left[(1+\lambda)(1-\lambda^2)\right] = \frac{1}{2} - \frac{\lambda^2}{2} + \frac{\lambda}{2} - \frac{\lambda^3}{2}$$

Set the derivative of rotor efficiency to zero and solve for λ:

$$\frac{\partial C_P}{\partial \lambda} = -2\lambda + 1 - 3\lambda^2 = 0$$

$$\frac{\partial C_P}{\partial \lambda} = 3\lambda^2 + 2\lambda - 1 = 0$$

$$\frac{\partial C_P}{\partial \lambda} = (3\lambda - 1)(\lambda + 1) = 0 \quad \longrightarrow \quad \lambda = 1/3$$

Maximizes rotor efficiency

When we plug the optimal value for λ back into C_P to find the maximum rotor efficiency:

$$C_P = \frac{1}{2}\left[(1 + 1/3)(1 - 1/3)\right] = 16/27 = 59.3\%. \tag{3.13}$$

The maximum efficiency of 59.3% occurs when air is slowed to one-third of its upstream rate. Again, this factor and value is called the "Betz efficiency" or "Betz' law" (Betz 1966). In plain English, Betz' law states that all wind power cannot be captured by the rotor; otherwise, air would be completely still behind the rotor and not allow more wind to pass through. For illustrative purposes, in Table 3.2, we list wind speed, power of the wind, and power of the wind based on Betz limit (59.3%).

TIP-SPEED RATIO (TSR)

Efficiency is a function of how fast the rotor turns. The tip-speed ratio (TSR) is an extremely important factor in wind turbine design. Tip-speed ratio is the ratio of the speed of the rotating blade tip to the speed of the free stream wind (see Figure 3.7). Stated differently, TSR is the speed of the outer tip of the blade divided by windspeed. There is an optimum angle of attack, which creates the highest lift to drag ratio. If the rotor of the wind turbine spins too slowly, most of the wind will pass straight through the gap between the blades, therefore giving it no power! But if the rotor spins too fast, the blades will blur and act like a solid wall to the wind. Moreover, rotor blades create turbulence as they spin through the air. If the next blade arrives too quickly, it will hit that turbulent air. Thus, it is better to slow down the blades. Because the angle of attack is dependent on wind speed, there is an optimum tip-speed ratio:

TABLE 3.2
Betz Limit for 80 M Rotor Turbine

Wind Speed mph/ms	Power (kW) of Wind	Power (kW) Betz Limit
5/2.2	36	21
10/4.5	285	169
15/6.7	962	570
20/8.9	2,280	1,352
25/11.2	4,453	2,641
28/12.5	6,257	3,710
30/13.4	7,695	4,563
35/15.6	12,220	7,246
40/17.9	18,241	10,817
45/20.1	25,972	15,401
50/22.4	35,626	21,126
55/24.6	47,419	28,,119
*56/25.0	50,053	29681
60/26.8	61,563	36,507

Source: Adapted from Devlin, L, 2007. *Wind Turbine Efficiency.* Accessed 12/12/21 @ http://k0lee.com/2007/11/wind-turbine-efficiency/

$$TSR = \Omega R/V, \qquad (3.14)$$

where
Ω = rotational speed in radians/sec,
R = Rotor Radius, and
V = Wind "Free Stream" Velocity.

SMALL-SCALE WIND POWER

In meeting your needs for electricity, a small home-size wind machine may be the answer. A small home-size wind turbine has rotors between eight and 25 feet in diameter and stands upward of 30 feet and can supply the power needs of an all-electric home or small business. Utility-scale turbines range in size from 50 to 750 kilowatts. Single small turbines, below 50 kilowatts, are used for homes, telecommunication dishes, or water pumping.

According to EERE (2006a), there are many questions potential users of small home-size wind turbines should ask and answer before installing a small wind system operation. In this section we provide some of the answers and explanations EERE provides in addressing these questions.

We begin by asking the obvious question: What are the benefits to homeowners from using wind turbines? Wind energy systems provide a cushion against electricity price increases. Wind energy systems reduce US dependence on fossil fuels, and they don't emit greenhouse gases. If you are building a home in a remote location, a small

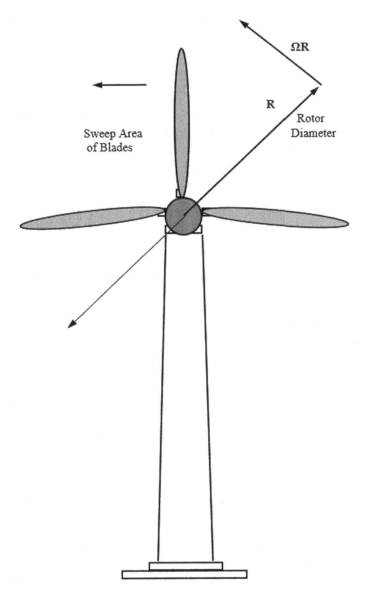

ΩR

R

Rotor
Diameter

Sweep Area
of Blades

FIGURE 3.7   Tip-speed ratio (TSR).

wind energy system can help you avoid the high cost of extending utility power lines
to your site.

Although wind energy systems involve a significant initial investment, they can
be competitive with conventional energy sources when you account for a lifetime
of reduced or altogether avoided utility costs. The length of payback time—the time
before the savings resulting from your system equal the system cost—depends on the

system you choose, the wind resource in your site, electric utility rates in your area, and how you use your wind system.

DID YOU KNOW?

The difference between power and energy is that power (kilowatts [kW]) is the rate at which electricity is consumed, while energy (kilowatt-hours [kWh]) is the quantity consumed.

Another frequently asked question: Is wind power practical for me? Small wind energy systems can be used in connection with an electricity transmission and distribution system (called a grid-connected system), or in stand-alone applications that are not connected to the utility grid. A grid-connected wind turbine can reduce your consumption of utility-supplied electricity for lighting, appliances, and electric heat. If the turbine cannot deliver the amount of energy you need, the utility makes up the difference. When the wind system produces more electricity than the household requires, the excess can be sold to the utility. With the interconnections available today, switching takes place automatically. Stand-alone wind energy systems can be appropriate for homes, farms, or even entire communities (a co-housing project, for example) that are far from the nearest utility lines.

Stand-alone systems (systems not connected to the utility gird) require batteries to store excess power generated for use when the wind is calm. They also need a charge controller to keep the batteries from overcharging. As mentioned earlier, deep-cycle batteries, such as those used for golf carts, can discharge and recharge 80% of their capacity hundreds of times, which makes them a good option for remote renewable energy systems. Automotive batteries are shallow-cycle batteries and should not be used in renewable energy systems because of their short life in deep-cycling operations.

Small wind turbines generate direct current (DC) electricity. In very small systems, DC appliances operate directly off the batteries. If you want to use standard appliances that use conventional household alternating current (AC), you must install an inverter to convert DC electricity from the batteries to AC (see Figure 3.8). Although the inverter slightly lowers the overall efficiency of the system, it allows the home to be wired for AC, a definite plus with lenders, electrical code officials, and future homebuyers.

SAFETY NOTE

For safety, batteries should be isolated from living areas and electronics because they contain corrosive and explosive substances. Lead-acid batteries also require protection from temperature extremes.

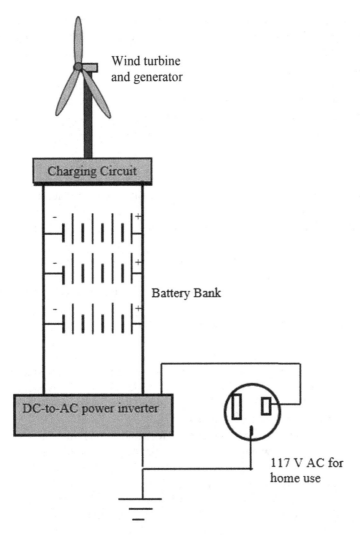

FIGURE 3.8   Stand-alone small-scale wind-power system.

In grid-connected systems (or interactive systems), the only additional equipment required is a power condition unit (inverter) that makes the turbine output electrically compatible with the utility grid. Usually, batteries are not needed.

Type of system can be practical if the following conditions exist.

CONDITIONS FOR STAND-ALONE SYSTEMS

- You live in an area with average annual wind speeds of at least 4.0 meters per second (9 mph)

- A grid connection is not available or can only be made through an expensive extension. The cost of running a power line to a remote site to connect with the utility grid can be prohibitive, ranging from $15,000 to more than $50,000 per mile, depending on terrain.
- You have an interest in gaining energy independence from the utility.
- You would like to reduce the environmental impact of electricity production.
- You acknowledge the intermittent nature of wind power and have a strategy for using intermittent resources to meet your power needs.

Conditions for Grid-Connected Systems

- You live in an area with average annual wind speeds of at least 4.5 meters per second (10 miles per hour).
- Utility-supplied electricity is expensive in your area (about 10 to 15 cents per kilowatt-hour).
- The utility's requirements for connecting your system to its grid are not prohibitively expensive.
- Local building codes or covenants allow you to legally erect a wind turbine on your property.
- You are comfortable with long-term investments.

After comparing stand-alone systems and grid-connected systems and determining which is best-suited for your circumstance, the next question to consider is whether your location is the right site to install a small-scale wind turbine system. That is, is there enough wind on your site? Is it legal to install the system on your property? Are there environmental and/or economic issues?

In determining whether your site is right for wind turbine installation, it must be determined whether the wind blows hard enough at your site to make a small wind turbine system economically worthwhile. That is a key question and not always easily answered. The wind resource can vary significantly over an area of just a few miles because of local terrain influences on the wind flow. Yet, there are steps you can take that will go a long way toward answering the above question.

As a first step, wind resource maps like the ones included in the USDOE's *Wind Energy Resource Atlas of United States* (RREDC, 2010) can be used to estimate the wind resource in your region. The highest average wind speeds in the United States are generally found along seacoasts, on ridgelines, and on the Great Plains; however, many areas have wind resources strong enough to power a small wind turbine economically. The wind resource estimates on the maps in the *Wind Energy Resource Atlas* generally apply to terrain features that are well exposed to the wind, such as plains, hilltops, and ridge crests. Local terrain features may cause the wind resource at a specific site to differ considerably from these estimates.

Average wind speed information can be obtained from a nearby airport. However, caution should be used because local terrain influences and other factors may cause the wind speed recorded at an airport to be different from your location. Airport wind data are generally measured at heights about 20–33 ft (6–10 m) above ground. Average wind speeds increase with height and may be 15%–25% greater at a typical wind turbine hub-height of 80 ft (24 m) than those measured at airport anemometer heights. The *Wind Energy Resource Atlas* contains data from airports in the United States and makes wind data summaries available.

Again, it is important to have site-specific data to determine the wind resource at your exact location. If you do not have on-site data and want to obtain a clearer, more predictable picture of your wind resource, you may wish to measure wind speeds at your location for a year. You can do this with a recording anemometer, which generally costs $500 to $1,500. The most accurate readings are taken at "hub height" (i.e., the elevation at the top of the wind turbine tower). This requires placing the anemometer high enough to avoid turbulence created by trees, buildings, and other obstructions. The standard wind sensor height used to obtain data for the USDOE maps is 10 meters (33 feet).

Within the same property it is not unusual to have varied wind resources. If you live in complex terrain, take care in selection of the installation site. If you site your wind turbine on the top or on the windy side of a hill, for example, you will have more access to prevailing wind than in a gully or on the leeward (sheltered) side of a hill on the same property. Consider existing obstacles and plan for future obstructions, including tees and buildings, which could block the wind. Also recall that the power in the wind is proportional to its speed (velocity) cubed (v^3). This means that the amount of power you get from your generator goes up exponentially as the wind speed increases. For example, if your site has an annual average wind speed of about 5.6 meters per second (12.6 miles per hour), it has twice the energy available as a site with a 4.5-meter-per-second (10 mile per hour) average ($12.6/10^3$).

Another useful indirect measurement of the wind resource is the observation of an area's vegetation. Trees, especially conifers or evergreens, can be permanently deformed by strong winds. The deformity, known as "flagging," has been used to estimate the average wind speed for an area (See Figure 3.9).

In addition to ensuring the proper siting of your small wind turbine system, there are also legal, environmental, and economic issues that must be addressed. For example, you should also

- research potential legal and environmental obstacles;
- obtain cost and performance information from manufacturers;
- perform a complete economic analysis that accounts for a multitude of factors;
- understand the basics of a small wind system; and
- review possibilities for combining your system with other energy sources, backups, and energy efficiency improvements.

Regarding economic issues, because energy efficiency is usually less expensive than energy production, making your house more energy efficient first will likely result in your being able to spend less money since you may need a smaller wind turbine to meet your needs. But, a word of caution. Before you spend any money, research potential legal and environmental obstacles to installing a wind system. Some jurisdictions, for example, restrict the height of the structures permitted in residentially zoned areas, although variances are often obtainable. Your neighbors might object to a wind machine that blocks their view, or they might be concerned about

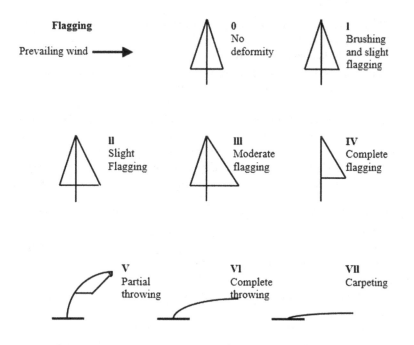

Griggs-Putnam Index of Deformity

Index	1	ll	lll	IV	V	VI	Vll
Wind mph	7-9	9-11	11-13	13-16	15-18	16-21	22+
Speed m/s	3-4	4-5	5-6	6-7	7-8	8-9	10

FIGURE 3.9 A crude method of approximating average annual wind speed from the deformation of trees (and other foliage).

noise. Consider obstacles that might block the wind in the future (large, planned developments or saplings, for example). Saplings that will grow into large trees can be a problem in the future. As mentioned, trees can affect wind speed (see Figure 3.9). If you plan to connect the wind generator to your local utility company's grid, find out its requirements for interconnections and about buying electricity from small independent power producers.

When you are convinced that a small wind turbine is what you want and there are no obstructions restricting its installation, approach buying a wind system as you would any major purchase.

THE BOTTOM LINE ON WIND POWER

According to Archer and Jacobson (2004), technology is much more advanced today in utilizing our wind resource, and the United States is home to one of the best wind resource areas in the world, the Midwest states of North and South Dakota, Nebraska, Kansas, Montana, Iowa, and Oklahoma. However, as with any other source of energy, nonrenewable or renewable, there are advantages and disadvantages associated with their use. On the positive side, it should be noted that wind energy is a free, renewable resource, so no matter how much is used today, there will still be the same supply in the future. Wind energy is also a source of clean, non-polluting, electricity. Wind turbines can be installed on farms or ranches, thus benefiting the economy in rural areas, where most of the best wind sites are found. Moreover, farmers and ranchers can continue to work the land because the wind turbines use only a fraction of the land—the height and distance between turbines means that land used for wind turbines can also be used for agriculture and grazing. That is, only about 5% of the land in a wind farm is occupied by the turbines themselves. One huge advantage of wind energy is that it is a domestic source of energy, produced in the United States or country where installed and where wind is abundant. Again, in the United States , the wind supply is abundant.

On the other side of the coin, wind energy does have a few negatives—wind projects face opposition. Wind power must compete with conventional generation sources on a cost basis. Even though the cost of wind power has decreased dramatically in the past 10 years, the technology requires a higher initial investment than fossil-fueled generators. The challenge to using wind as a source of power is that the wind is intermittent and it does not always blow when electricity is needed. Wind energy cannot be stored (unless batteries are being used), and not all winds can be harnessed to meet the timing of electricity demands. Another problem is that good sites are often located in remote locations, far from cities where the electricity is needed. Moreover, wind resource development may compete with other uses for the land and those alternative uses may be more highly valued than electricity generation. Finally, regarding the environment, wind power plants have relatively little impact on the environment compared to other conventional power plants. However, there is some concern about the noise produced by the rotor blades, aesthetic (visual) impacts, and sometimes birds have been killed by flying into the rotors. Most of these problems have been resolved or greatly reduced through technological development or by properly siting wind plants. But keep in mind that the "not in my backyard" (NIMBY) point of view is still alive and strong in the United States and has succeeded in killing many projects, even renewable, "clean" energy projects.

We understand some of the issues related to the NIMBY point of view, although we don't agree with them. While it is questionable and arguable that the United States has reached its apogee in attaining the so-called good life for most of its citizenry and that many feel that our best days are behind us, it is not questionable about our future if we do not innovate and free ourselves from the bondage of fossil fuels and the accompanying economics, politics, and potential turmoil. With fossil fuel supplies dwindling and world political pressures for accessibility ramping up to dangerous levels, we simply need to find reliable alternative energy sources for the future. We may never be able to improve on the so-called "good life," but it would be beneficial for most of us if we are able to maintain the level of living that we have obtained to this point while at the same time reducing oil imports and carbon emissions.

One thing is certain; wind energy production continues to be one of the fastest growing energy technologies; it looks set to become a major generator of electricity throughout the world.

NOTES

1   Parts of this section are adapted from F.R. Spellman and N.E. Whiting 2006 *Environmental Science and Technology*. Lanham, MD: Government Institutes Press.
2   Much of the information in this section is from USDOE-EERE 2005. *History of wind energy*. Accessed 06/14/09 @ www1/eere/emergu/gpv/womdamdjudrp/printable_versions/wind_hisotry.htm

REFERENCES AND RECOMMENDED READING

Archer, C. and Jacobson, M.Z. 2004. *Evaluation of Global Wind Power*. Stanford, CA: Department of Civil and Environmental Engineering, Stanford University.

Betz, A. 1966. *Introduction to the Theory of Flow Machines*. (D.G. Randall, Trans.) Oxford: Pergamon Press.

Energy Efficiency & Renewable Energy (EERE). 2006a. *Wind & Hydropower Technologies Program*. Accessed 02/20/10 @ www1.3343.energy.gov/windandhydro/wind_how.html

Hanson, B.J. 2004. *Energy Power Shift*. Maple, WI: Lakota Scientific Press.

Khaaligh, A. and Onar, O.C. 2010. *Energy Harvesting*. Boca Raton, FL: Taylor and Francis.

Redwood Region Economic Development Commission (RREDC). 2010. *Wind Energy Resource Atlas of the US*. Accessed 02/25/23 @ rredc.nrel.gov/Wind/pubs/atlas

WINDEIS. 2009. *Wind Energy Guide*. Accessed 06/13/2023 @ http://windeis.anl.gov/guide/basics/index.cfm

4 Hydropower

When we speak of water, and its many manifestations we are speaking of that endless quintessential cycle that predates all other cycles. Water is our most precious natural resource; we can't survive without it. There is no more water today than there was yesterday; that is, no more this calendar year than 100 million years ago. The water present today is the same water used by all the animals that ever lived, by cave dwellers, Caesar, Cleopatra, Christ, da Vinci, John Snow, Teddy Roosevelt, and the rest of us—again there is not one drop more or one drop less of water than there has always been. This life-giving cycle, though it is a unique blend of thermal and mechanical aspects, is dependent on solar energy and gravity for its existence (see Figure 4.1). Nothing on earth is truly infinite in supply, but the energy available from water sources, in practical terms, comes closest to that ideal.

–F.R. Spellman (2008)

INTRODUCTION

When we look at rushing waterfalls and rivers, we may not immediately think of electricity. But hydroelectric (water-powered) power plants are responsible for lighting many of our homes and neighborhoods. *Hydropower* is the harnessing of water to perform work. The power of falling water has been used in industry for thousands of years (see Table 4.1). The Greeks used water wheels for grinding wheat into flour more than 2,000 years ago. Besides grinding flour, the power of the water was used to saw wood and power textile mills and manufacturing plants.

Hydropower (2018) reports that in 2017 alone, it is estimated that 4 billion tons of greenhouse gases were not released into our atmosphere by generating electricity from hydropower.

The technology for using falling water to create hydroelectricity has existed for more than a century. The evolution of the modern hydropower turbine began in the mid-1700s when a French hydraulic and military engineer, Bernard Forest de Belidor, wrote a four-volume work describing the use of a vertical-axis versus a horizontal-axis machine.

DOI: 10.1201/9781003439059-5

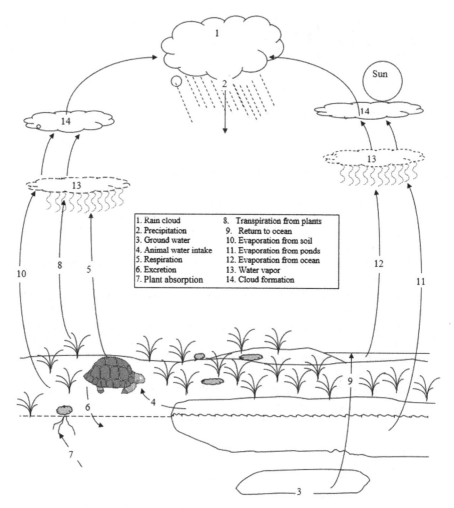

FIGURE 4.1   Water cycle.

Source: Modified from Carolina Biological Supply Co., 1966. The Water Cycle. Burlington, NC.

Water turbine development continued during the 1700s and 1800s. In 1880, a brush arc light dynamo driven by a water turbine was used to provide theater and storefront lighting in Grand Rapids, Michigan; and in 1881, a brush dynamo connected to a turbine in a flour mill provided street lighting at Niagara Falls, New York. These two projects used direct-current (DC) technology.

Alternating current (AC) is used today. That breakthrough came when the electric generator was coupled to the turbine, which resulted in the world's, and the Unites States', first hydroelectric plant located on the Fox River in Appleton, Wisconsin, in 1882. The US Library of Congress (2009) lists the Appleton hydroelectric power

TABLE 4.1
History of Hydropower

Date	Hydropower Event
BCE	Hydropower used by the Greeks to turn water wheels for grinding wheat into flour, more than 2,000 years ago.
Mid-1770s	French hydraulic and military engineer Bernard Forest de Belidor wrote a four-volume work describing vertical- and horizontal-axis machines.
1775	US Army Corps of Engineers founded, with establishment of Chief Engineer for the Continental Army.
1880	Michigan's Grand Rapids Electric Light and Power Company, generating electricity by dynamo belted to a water turbine at the Wolverine Chair Factory, lit up 16 brush-arc lamps
1881	Niagara Falls city street amps powered by hydropower.
1882	World's first hydroelectric power plant began operation on the Fox River in Appleton, Wisconsin.
1886	About 45 water-powered electric plants in the United States and Canada.
1887	San Bernardino, CA, opens first hydroelectric plant in the West.
1889	Two hundred electric plants in the United States use waterpower for some or all generation.
1901	First Federal Water Power Act.
1902	Bureau of Reclamation established.
1907	Hydropower provided 15% of US electrical generation.
1920	Hydropower provided 25% of US electrical generation. Federal Power Act establishes Federal Power Commission authority to issue licenses for hydro development on public lands.
1933	Tennessee Valley Authority established.
1935	Federal Power Commission authority extended to all hydroelectric projects built by utilities engaged in interstate commerce.
1937	Bonneville Dam, first Federal dam, begins operation on the Columbia River; Bonneville Power Administration established.
1940	Hydropower provided 40% of electrical generation. Conventional capacity tripled in United States since 1920.
1980	Conventional capacity nearly tripled in United States since 1900.
2003	About 10% of US electricity comes from hydropower. Today, there is about 80,000MW of conventional capacity and 18,000 MW of pumped storage.

Source: EERE (2008).

plant as one of the major accomplishments of the Gilded Age (1878–1889). Soon, people across the United States were enjoying electricity in homes, schools, and offices, reading by electric lamp instead of candlelight or kerosene. Today, we take electricity for granted, not able to imagine life without it.

Ranging in size from small systems (100 kilowatts to 30 megawatts) for a home or village, to large projects (capacity greater than 30 megawatts) producing electricity for utilities, hydropower plants are of three types: impoundment, diversion, and pumped storage. Some hydropower plants use dams and some do not. Many dams were built for other purposes and hydropower was added later. In the United States, there are about 80,000 dams, of which only 2,400 produce power. The other dams are for recreation, stock/farm ponds, flood control, water supply, and irrigation. The types of hydropower plants are described below.

IMPOUNDMENT

The most common type of hydroelectric power plant is an impoundment facility. An impoundment facility, typically a large hydropower system, uses a dam to store river water in a reservoir. This type of facility works best in mountainous or hilly terrain where high dams can be built and deep reservoirs can be maintained. Potential energy available in a reservoir depends on the mass of water contained in it, as well as on overall depth of the water. Water released from the reservoir flows through a turbine, spinning it, which in turn activates a generator to produce electricity. The water may be released either to meet changing electricity needs or to maintain a constant reservoir level.

DID YOU KNOW?

The potential energy in a specific slug or parcel of water is expressed in newton-meters (N · m). The newton is the standard unit of force, equivalent to a meter per second squared (1 m/s^2). This is the product of the mass of the slug in kilograms, the acceleration of gravity in meters per second squared (about 9.8 m/s^2), and the elevation of the parcel in meters (the vertical distance it falls as its energy is harnessed). The equivalent kinetic-energy unit is the joule (J), which is in effect equal to a watt-second (W · s).

DIVERSION

A diversion, sometimes called run-of-river, facility channels all or a portion of the flow of a river from its natural course through a canal or penstock, and the current through this medium is used to drive turbine. It may not require the use of a dam. This type of system is best suited for locations where a river drops considerably per unit of horizontal distance. The ideal location is near a natural waterfall or rapids. The chief advantage of a diversion system is the fact that, lacking a dam, it has far less impact on the environment than an impoundment facility (Gibilisco. 2007).

PUMPED STORAGE HYDROPOWER "BATTERY"

Pumped stored hydropower is a type of hydroelectric energy storage. Basically, it is a configuration of two water reservoirs at different elevations that can generate power

TABLE 4.2
US Energy Consumption by Energy Source, 2022 (Quadrillion Btu and %)

Energy Source	2022
Total	**97.33**
Renewable	12.16
Biomass (biofuels, waste, wood and wood-derived)	40%
Biofuels	19%
Waste	4%
Wood-derived Fuels	17%
Geothermal	2%
Hydroelectric Conventional	19%
Solar/PV	12%
Wind	27%

Source: EIA 2022. *US Energy Consumption by Energy Source.* Accessed 06/25/23 @ www.eia.energyex plained/us-energy-facts.

as water moves down (discharges) from one to the other, passing through a turbine. The system also requires power as it recharges by pumping water back into the upper reservoir. When the demand for electricity is low, a pumped storage facility stores energy by pumping water from a lower reservoir to an upper reservoir. During periods of high electrical demand, the water is released back to the lower reservoir to generate electricity. In a sense, pumped storage hydropower is like a giant battery, because it can store power and then release it when needed.

Table 4.2 highlights hydropower energy's percentage ranking in current renewable energy source use. As pointed out in the table, the energy consumption by energy source computations are from the year 2022. Thus, the 19% quantity is expected to steadily increase to a higher level and this should be reflected in the 2022–2023 figures when they are released.

HYDROPOWER BASIC CONCEPTS[1]

Air Pressure (@ Sea Level) = 14.7 pounds per square inch (psi)

The relationship shown above is important because our study of hydropower basics begins with air. A blanket of air, many miles thick, surrounds the Earth. The weight of this blanket on a given square inch of the Earth's surface will vary according to the thickness of the atmospheric blanket above that point. As shown above, at sea level, the pressure exerted is 14.7 pounds per square inch (psi). On a mountaintop, air pressure decreases because the blanket is not as thick.

$$1 \text{ ft}^3 \text{ H}_2\text{O} = 62.4 \text{ lb}$$

The relationship shown above is also important: both cubic feet and pounds are used to describe a volume of water. There is a defined relationship between these two methods of measurement. The specific weight of water is defined relative to a cubic foot. One cubic foot of water weighs 62.4 pounds. This relationship is true only at a temperature of 4°C and at a pressure of one atmosphere (known as standard temperature and pressure (STP)—14.7 lbs. per square inch at sea level containing 7.48 gallons). The weight varies so little that, for practical purposes, this weight is used from a temperature 0°C to 100°C. One cubic inch of water weighs 0.0362 pounds. Water one foot deep will exert a pressure of 0.43 pounds per square inch on the bottom area (12 in x 0.0362 lb/in³). A column of water two feet high exerts 0.86 psi, one 10 feet high exerts 4.3 psi, and one 55 feet high exerts

$$55 \text{ ft x } 0.43 \text{ psi/ft} = 23.65 \text{ psi.}$$

A column of water 2.31 feet high will exert 1.0 psi. To produce a pressure of 50 psi requires a water column

$$50 \text{ psi x } 2.31 \text{ ft/psi} = 115.5 \text{ ft.}$$

Remember: The important points being made here are:

1. $1 \text{ ft}^3 \text{ H}_2\text{O} = 62.4 \text{ lb}$
2. A column of water 2.31 ft high will exert 1.0 psi

Another relationship is also important:

$$1 \text{ gallon H}_2\text{O} = 8.34 \text{ pounds.}$$

At standard temperature and pressure, one cubic foot of water contains 7.48 gallons. With these two relationships, we can determine the weight of one gallon of water. This is accomplished by

$$\text{wt. of gallon of water} = 62.4 \text{ lb} \div 7.48 \text{ gal} = 8.34 \text{ lb/gal.}$$

Thus,

$$1 \text{ gallon H}_2\text{O} = 8.34 \text{ pounds.}$$

Note: Further, this information allows cubic feet to be converted to gallons by simply multiplying the number of cubic feet by 7.48 gal/ft³.

EXAMPLE 4.1

Problem:

Find the number of gallons in a reservoir that has a volume of 855.5 ft³.

Solution:

$$855.5 \text{ ft}^3 \text{ x } 7.48 \text{ gal/ft}^3 = 6,399 \text{ gallons (rounded)}$$

Note: The term "head" is discussed later, but for now it is important to point out that it is used to designate water pressure in terms of the height of a column of water in feet. For example, a 10-foot column of water exerts 4.3 psi. This can be called 4.3-psi pressure or 10 feet of head.

STEVIN'S LAW

Stevin's Law deals with water at rest. Specifically, the Law states: "The pressure at any point in a fluid at rest depends on the distance measured vertically to the free surface and the density of the fluid." Stated as a formula, this becomes

$$p = w \text{ x } h, \quad\quad\quad\quad (4.1)$$

where
   p = pressure in pounds per square foot (psf),
   w = density in pounds per cubic foot (lb/ft^3), and
   h = vertical distance in feet.

EXAMPLE 4.2

Problem:

What is the pressure at a point 18 feet below the surface of a reservoir?

Solution:

Note: To calculate this, we must know that the density of the water, w, is 62.4 pounds per cubic foot.

$$p = w \text{ x } h$$

$$= 62.4 \text{ lb/ft}^3 \text{ x } 18 \text{ ft}$$

$$= 1123 \text{ lb/ft}^2 \text{ or } 1123 \text{ psf}$$

Water practitioners generally measure pressure in pounds per square *inch* rather than pounds per square *foot*; to convert, divide by 144 in^2/ft^2 (12 in x 12 in = 144 in^2):

$$P = \frac{1123\,\text{psf}}{144 \text{ in}^2 / \text{ft}^2} = 7.8\,\text{lb} / \text{in}^2 \text{ or psi}\left(\text{rounded}\right).$$

PROPERTIES OF WATER

Table 4.3 shows the relationship between temperature, specific weight, and density of water.

DENSITY AND SPECIFIC GRAVITY

When we say that iron is heavier than aluminum, we say that iron has greater density than aluminum. In practice, what we are really saying is that a given volume of iron is heavier than the same volume of aluminum.

Note: What is density? *Density* is the *mass per unit volume* of a substance.

Suppose you had a tub of lard and a large box of cold cereal, each having a mass of 600 grams. The density of the cereal would be much less than the density of the lard because the cereal occupies a much larger volume than the lard occupies. The density of an object can be calculated by using the formula:

$$\text{Density} = \frac{\text{Mass}}{\text{Volume}}. \tag{4.2}$$

Perhaps the most common measures of density are pounds per cubic foot (lb/ft^3) and pounds per gallon (lb/gal):

- *1 cu ft of water weight 62.4 lbs—Density = 62.4 lb/cu/ft*
- *One gallon of water weighs 8.34 lbs—Density = 8.34 lb/gal*

The density of a dry material, such as cereal, lime, soda, and sand, is usually expressed in pounds per cubic foot. The density of a liquid, such as liquid alum, liquid

TABLE 4.3
Water Properties (Temperature, Specific Weight, and Density)

Temperature (°F)	Specific Weight (lb/ft³)	Density (slugs/ft³)	Temperature (°F)	Specific Weight (lb/ft³)	Density (slugs/ft³)
32	62.4	1.94	130	61.5	1.91
40	62.4	1.94	140	61.4	1.91
50	62.4	1.94	150	61.2	1.90
60	62.4	1.94	160	61.0	1.90
70	62.3	1.94	170	60.8	1.89
80	62.2	1.93	180	60.6	1.88
90	62.1	1.93	190	60.4	1.88
100	62.0	1.93	200	60.1	1.87
110	61.9	1.92	210	59.8	1.86
120	61.7	1.92			

chlorine, or water, can be expressed either as pounds per cubic foot or as pounds per gallon. The density of a gas, such as chlorine gas, methane, carbon dioxide, or air, is usually expressed in pounds per cubic foot.

As shown in Table 4.3, the density of a substance like water changes slightly as the temperature of the substance changes. This occurs because substances usually increase in volume (size—they expand) as they become warmer. Because of this expansion with warming, the same weight is spread over a larger volume, so the density is lower when a substance is warm than when it is cold.

Note: What is specific gravity? Specific gravity is the weight (or density) of a substance compared to the weight (or density) of an equal volume of water. (*Note:* The specific gravity of water is 1).

This relationship is easily seen when a cubic foot of water, which weighs 62.4 lbs as shown earlier, is compared to a cubic foot of aluminum, which weighs 178 pounds. Aluminum is 2.7 times as heavy as water. It is not that difficult to find the specific gravity of a piece of metal. All you must do is weigh the metal in air, then weigh it under water. Its loss of weight is the weight of an equal volume of water. To find the specific gravity, divide the weight of the metal by its loss of weight in water.

$$\text{Sepcific Gravity} = \frac{\text{Weight of Substance}}{\text{Weight of Equal Volume of Water}}. \qquad (4.3)$$

EXAMPLE 4.3

Problem:

Suppose a piece of metal weighs 150 pounds in air and 85 pounds under water. What is the specific gravity?

Solution:

Step 1: 150 lb subtract 85 lb = 65 lb loss of weight in water

Step 2:

$$\text{specific gravity} = \frac{150}{65} = 2.3$$

DID YOU KNOW?

In a calculation of specific gravity, it is *essential* that the densities be expressed in the same units.

As stated earlier, the specific gravity of water is one (1), which is the standard, the reference to which all other liquid or solid substances are compared. Specifically, any object that has a specific gravity greater than one will sink in water (rocks, steel, iron, grit, floc, sludge). Substances with a specific gravity of less than one will float (wood, scum, gasoline). Considering the total weight and volume of a ship, its specific gravity is less than one; therefore, it can float.

The most common use of specific gravity in water operations is in gallons-to-pounds conversions. In many cases, the liquids being handled have a specific gravity of 1.00 or very nearly 1.00 (between 0.98 and 1.02), so 1.00 may be used in the calculations without introducing significant error. However, in calculations involving a liquid with a specific gravity of less than 0.98 or greater than 1.02, the conversions from gallons to pounds must consider specific gravity. The technique is illustrated in the following example.

EXAMPLE 4.4

Problem:

There are 1,455 gal of a certain liquid in a basin. If the specific gravity of the liquid is 0.94, how many pounds of liquid are in the basin?

Solution:

Normally, for a conversion from gallons to pounds, we would use the factor 8.34 lb/gal (the density of water) if the substance's specific gravity were between 0.98 and 1.02. However, in this instance the substance has a specific gravity outside this range, so the 8.34 factor must be adjusted.

Multiply 8.34 lb/gal by the specific gravity to obtain the adjusted factor:

Step 1: (8.34 lb/gal) (0.94) = 7.84 lb/gal (rounded)
Step 2: Then convert 1,455 gal to pounds using the corrected factor:

$$(1,455 \text{ gal}) (7.84 \text{ lb/gal}) = 11,407 \text{ lb (rounded)}$$

FORCE AND PRESSURE

Water exerts force and pressure against the walls of its container, whether it is stored in a tank or flowing in a pipeline. There is a difference between force and pressure, though they are closely related. Force and pressure are defined below.

Force is the push or pull influence that causes motion. In the English system, force and weight are often used in the same way. The weight of a cubic foot of water is 62.4 pounds. The force exerted on the bottom of a one-foot cube is 62.4

pounds. If we stack two cubes on top of one another, the force on the bottom will be 124.8 pounds.

Pressure is a force per unit of area. In equation form, this can be expressed as:

$$P = \frac{F}{A},\qquad\qquad (4.4)$$

where

P = pressure,

F = force, and

A = area over which the force is distributed.

Earlier we pointed out that pounds per square inch or pounds per square foot are common expressions of pressure. The pressure on the bottom of the cube is 62.4 pounds per square foot. It is normal to express pressure in pounds per square inch (psi). This is easily accomplished by determining the weight of one square inch of a cube one foot high. If we have a cube that is 12 inches on each side, the number of square inches on the bottom surface of the cube is 12 x 12 = 144 in². Dividing the weight by the number of square inches determines the weight on each square inch:

$$\mathrm{psi} = \frac{62.4 \text{ lb} / \text{ft}}{144 \text{ in}^2} = 0.433\mathrm{psi} / \text{ft}$$

This is the weight of a column of water one-inch square and one foot tall. If the column of water were two feet tall, the pressure would be 2 ft x 0.433 psi/ft = 0.866.

1 foot of water = 0.433 psi

With the above information, feet of head can be converted to psi by multiplying the feet of head times 0.433 psi/ft.

EXAMPLE 4.5

Problem:

A tank is mounted at a height of 90 feet. Find the pressure at the bottom of the tank.

Solution:

90 ft x 0.433 psi/ft = 39 psi (rounded)

To convert psi to feet, you would divide the psi by 0.433 psi/ft.

EXAMPLE 4.6

Problem:

Find the height of water in a tank if the pressure at the bottom of the tank is 22 psi.

Solution:

$$\text{height in feet} = \frac{22\ \text{psi}}{0.433\ \text{psi}/\text{ft}} = 51\text{ft}\,(\text{rounded})$$

DID YOU KNOW?

One of the problems encountered in a hydraulic system is storing the liquid. Unlike air, which is readily compressible and is capable of being stored in large quantities in relatively small containers, a liquid such as water cannot be compressed. Therefore, it is not possible to store a large amount of water in a small tank: 62.4 lbs of water occupies a volume of one cubic foot, regardless of the pressure applied to it.

HYDROSTATIC PRESSURE

Figure 4.2 shows several differently shaped, connected, open containers of water. Note that the water level is the same in each container, regardless of the shape or size of the container. This occurs because pressure is developed, within water (or any other liquid), by the weight of the water above. If the water level in any one container were to be momentarily higher than that in any of the other containers, the higher pressure at the bottom of this container would cause some water to flow into the container having the lower liquid level. In addition, the pressure of the water at any level (such as Line T) is the same in each of the containers. Pressure increases because

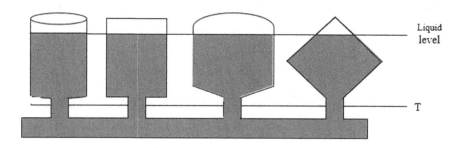

FIGURE 4.2   Hydrostatic pressure.

of the weight of the water. The farther down from the surface, the more pressure is created. This illustrates that the *weight*, not the volume, of water contained in a vessel determines the pressure at the bottom of the vessel.

Nathanson (1997) points out some very important principles that always apply for hydrostatic pressure:[23]

1. The pressure depends only on the depth of water above the point in question (not on the water surface area).
2. The pressure increases in direct proportion to the depth.
3. The pressure in a continuous volume of water is the same at all points that are at the same depth.
4. The pressure at any point in the water acts in all directions at the same depth.

HEAD

Head is defined as the vertical distance the water must be lifted from the supply tank to the discharge, or as the height a column of water would rise due to the pressure at its base. A perfect vacuum plus atmospheric pressure of 14.7 psi would lift the water 34 feet. If the top of the sealed tube is opened to the atmosphere and the reservoir is enclosed, the pressure in the reservoir is increased; the water will rise in the tube. Because atmospheric pressure is essentially universal, we usually ignore the first 14.7-psi of actual pressure measurements, and measure only the difference between the water pressure and the atmospheric pressure; we call this *gauge pressure*. For example, water in an open reservoir is subjected to the 14.7 psi of atmospheric pressure, but subtracting this 14.7 psi leaves a gauge pressure of 0 psi. This shows that the water would rise 0 feet above the reservoir surface. If the gauge pressure in a water main were 120 psi, the water would rise in a tube connected to the main:

$$120 \text{ psi} \times 2.31 \text{ ft/psi} = 277 \text{ ft (rounded)}.$$

The *total head* includes the vertical distance the liquid must be lifted (static head), the loss to friction (friction head), and the energy required to maintain the desired velocity (velocity head):

$$\text{Total Head} = \text{Static Head} + \text{Friction Head} + \text{Velocity Head}. \qquad (4.5)$$

- *Static Head*—is the actual *vertical* distance the liquid must be lifted:

$$\text{Static Head} = \text{Discharge Elevation} - \text{Supply Elevation}. \qquad (4.6)$$

EXAMPLE 4.7

Problem:

The supply tank is located at elevation 118 feet. The discharge point is at elevation 215 feet. What is the static head in feet?

Solution:

Static Head, ft = 215 ft – 118 ft = 97 feet

- *Friction Head—Friction head* is the equivalent distance of the energy that must be supplied to overcome friction. Engineering references include tables showing the equivalent vertical distance for various sizes and types of pipes, fittings, and valves. The total friction head is the sum of the equivalent vertical distances for each component.

Friction Head, ft = Energy Losses Due to Friction.   (4.7)

- *Velocity Head—Velocity head* is the equivalent distance of the energy consumed in achieving and maintaining the desired velocity in the system.

Velocity Head, ft = Energy Losses to Maintain Velocity.   (4.8)

- *Total Dynamic Head (Total System Head)*

Total Head = Static Head + Friction Head + Velocity Head   (4.9)

- *Pressure/Head*

The pressure exerted by water is directly proportional to its depth or head in the pipe, tank, or channel. If the pressure is known, the equivalent head can be calculated:

Head, ft = Pressure, psi x 2.31 ft/psi.   (4.10)

EXAMPLE 4.8

Problem:

The pressure gauge on the discharge line from the influent pump reads 72.3 psi. What is the equivalent head in feet?

Solution:

Head, ft = 72.3 x 2.31 ft/psi = 167 ft

- *Head/Pressure*

If the head is known, the equivalent pressure can be calculated by:

$$\text{Pressure, psi} = \frac{\text{Head, ft}}{2.31 \text{ ft / psi}}. \qquad (4.11)$$

EXAMPLE 4.9

Problem:

The tank is 22 feet deep. What is the pressure in psi at the bottom of the tank when it is filled with water?

Solution:

$$\text{Pr essure, psi} = \frac{22 \text{ ft}}{2.31 \text{ ft / psi}} = 9.52 \text{psi (rounded)}$$

FLOW/DISCHARGE RATE: WATER IN MOTION

The study of fluid flow is much more complicated than that of fluids at rest, but it is important to understand these principles because the water in used in hydropower applications is nearly always in motion (e.g., the water used to propel turbine blades).

Discharge (or flow) is the quantity of water passing a given point in a pipe or channel during a given period. Stated another way for open channels, the flow rate through an open channel is directly related to the velocity of the liquid and the cross-sectional area of the liquid in the channel:

$$Q = A \times V, \tag{4.12}$$

where
  Q = Flow – discharge in cubic feet per second (cfs),
  A = Cross-sectional area of the pipe or channel (ft²), and
  V = water velocity in feet per second (fps or ft/sec).

EXAMPLE 4.10

Problem:

The channel is 6 feet wide and the water depth is 3 feet. The velocity in the channel is 4 feet per second. What is the discharge or flow rate in cubic feet per second?

Solution:

Flow, cfs = 6 ft x 3 ft x 4 ft/second = 72 cfs.

Discharge or flow can be recorded as gallons/day (gpd), gallons/minute (gpm), or cubic feet (cfs). Flows treated by many hydropower systems are large, and often referred to in million gallons per day (MGD). The discharge or flow rate can be converted from cfs to other units such as gallons per minute (gpm) or million gallons per day (MGD) by using appropriate conversion factors.

EXAMPLE 4.11

Problem:

A pipe 12 inches in diameter has water flowing through it at 10 feet per second. What is the discharge in (a) cfs, (b) gpm, and (c) MGD? Before we can use the basic formula (4.13), we must determine the area A of the pipe. The formula for the area of a circle is

$$A = \pi \times \frac{D^2}{4} = \pi \times r^2 \qquad (4.13)$$

(π is the constant value 3.14159 or simply 3.14),

where
  D = diameter of the circle in feet and
  r = radius of the circle in feet.

Therefore, the area of the pipe is:

$$D^2 \ (1 \ \text{ft})^2$$

$$A = \pi \frac{D^2}{4} = 3.14 \times \frac{\left(1\text{ft}\right)^2}{4} = 0.785 \text{ft}^2.$$

Now we can determine the discharge in cfs [part (a)]:

$$Q = V \times A = 10 \ \text{ft/sec} \times 0.785 \ \text{ft}^2 = 7.85 \ \text{ft}^3/\text{sec or cfs.}$$

For part (b), we need to know that 1 cubic foot per second is 449 gallons per minute, so 7.85 cfs x 449 gpm/cfs = 3525 gpm. (rounded)

Finally, for part (c), one million gallons per day is 1.55 cfs, so

$$\frac{7.85 \ \text{cfs}}{1.55 \dfrac{\text{cfs}}{\text{MGD}}} = 5.06 \quad \text{MGD.}$$

AREA/VELOCITY

The *law of continuity* states that the discharge at each point in a pipe or channel is the same as the discharge at any other point (if water does not leave or enter the pipe or channel). That is, under the assumption of steady state flow, the flow that enters the pipe or channel is the same flow that exits the pipe or channel. In equation form, this becomes

$$Q_1 = Q_2 \text{ or } A_1V_1 = A_2V_2. \qquad (4.14)$$

EXAMPLE 4.12

Problem:

A pipe 12 inches in diameter is connected to a 6-inch diameter pipe. The velocity of the water in the 12-inch pipe is 3 fps. What is the velocity in the 6-inch pipe?

Solution:

Using the equation $A_1V_1 = A_2V_2$, we need to determine the area of each pipe:

$$12 \text{ in.}: A = \pi \times \frac{D^2}{4}$$

$$= 3.14 \times \frac{(1\,\text{ft})^2}{4}$$

$$= 0.785 \text{ ft}^2$$

$$6 \text{ in.}: A = 3.14 \times \frac{(0.5)^2}{4}$$

$$= 0.196 \text{ ft}^2.$$

The continuity equation now becomes

$$(0.785 \text{ ft}^2) \times \left(3\,\frac{\text{ft}}{\text{sec}}\right) = (0.196 \text{ ft}^2) \times V_2.$$

Solving for V_2,

$$V_2 = \frac{(0.785\,\text{ft}^2) \times (3\,\text{ft}/\text{sec})}{(0.196\,\text{ft}^2)}$$

$$= 12 \text{ ft/sec or fps.}$$

PRESSURE/VELOCITY

In a closed pipe flowing full (under pressure), the pressure is indirectly related to the velocity of the liquid:

$$\text{Velocity}_1 \text{ x Pressure}_1 = \text{Velocity}_2 \text{ x Pressure}_2 \qquad (4.15)$$

or

$$V_1 P_1 = V_2 P_2.$$

PIEZOMETRIC SURFACE AND BERNOULLI'S THEOREM

The volume of water flowing past any given point in the pipe or channel per unit time is called the flow rate or discharge—or just flow. With regard to flow, continuity of flow and the continuity equation have been discussed (i.e., Equation 4.15). Along with the continuity of flow principle and continuity equation, the law of conservation of energy, piezometric surface, and Bernoulli's theorem (or principle) are also important to our study of water hydraulics.

CONSERVATION OF ENERGY

Many of the principles of physics are important to the study of hydraulics. When applied to problems involving the flow of water, few of the principles of physical science are more important and useful to us than the *Law of Conservation of Energy*. Simply, the Law of Conservation of Energy states that energy can neither be created nor destroyed, but it can be converted from one form to another. In each closed system, the total energy is constant.

ENERGY HEAD

As previously mentioned, there are two types of energy, kinetic and potential. Three forms of mechanical energy exist in hydraulic systems: potential energy due to elevation, potential energy due to pressure, and kinetic energy due to velocity. Energy has the units of foot pounds (ft-lbs). It is convenient to express hydraulic energy in terms of *Energy Head*, in feet of water. This is equivalent to foot-pounds per pound of water (ft lb/lb = ft).

ENERGY AVAILABLE

Energy available is directly proportional to flow rate and to the hydraulic head. As mentioned, the head is equivalent to stored potential energy. This is shown as

$$= mgh,$$

where
  m = the mass of water,
  g = the acceleration due to gravity (can be taken as 10ms^{-2}) in most applications, and
  h = the head difference.

With regard to piping, the pipe diameter must be large enough to take the volume of water flowing. Friction in the pipes will reduce the effective head of water, and larger diameters are used, although cost then has a bearing. Ideally, the pipes should narrow as one proceeds downhill; however, friction losses are highest where the velocity is highest, and so there is usually little change in pipe diameter. Friction losses in piping are classified as either major head loss or minor head loss (Tovey, 2005).

MAJOR HEAD LOSS

Major head loss consists of pressure decreases along the length of pipe caused by friction created as water encounters the surfaces of the pipe. It typically accounts for most of the pressure drop in a pressurized or dynamic water system. The components that contribute to major head loss are the following: roughness, length, diameter, and velocity.

- *Roughness*—Even when new, the interior surfaces of pipes are rough. The roughness varies, of course, depending on pipe material, corrosion (tuberculation and pitting), and age. Because normal flow in a water pipe is turbulent, the turbulence increases with pipe roughness, which, in turn, causes pressure to drop over the length of the pipe.
- *Pipe Length*—With every foot of pipe length, friction losses occur. The longer the pipe, the more head loss. Friction loss because of pipe length must be factored into head loss calculations.
- *Pipe Diameter*—Generally, small diameter pipes have more head loss than large diameter pipes. This is the case because in large diameter pipes less water touches the interior surfaces of the pipe (encountering less friction) than in a small diameter pipe.
- *Water Velocity*—Turbulence in a water pipe is directly proportional to the speed (or velocity) of the flow. Thus, the velocity head also contributes to head loss.

DID YOU KNOW?

For the same diameter pipe, when flow increases and head loss increases.

Calculating Major Head Loss

Darcy, Weisbach, and others developed the first practical equation used to determine pipe friction in about 1850. The equation or formula now known as the *Darcy-Weisbach* equation for circular pipes is:

$$h_f = f\frac{LV^2}{D2g}. \qquad (4.16)$$

In terms of the flow rate Q, the equation becomes:

$$h_f = \frac{8fLQ^2}{\pi^2 gD^5}, \qquad (4.17)$$

where h_f = head loss, (ft)
  f = coefficient of friction,
  L = length of pipe, (ft)
  V = mean velocity, (ft/s)
  D = diameter of pipe, (ft)
  g = acceleration due to gravity, (32.2 ft/s^2) and
  Q = flow rate (ft^3/s).

The Darcy-Weisbach formula as such was meant to apply to the flow of any fluid and into this friction factor was incorporated the degree of roughness and an element called the *Reynold's Number*, which was based on the viscosity of the fluid and the degree of turbulence of flow.

The Darcy-Weisbach formula is used primarily for determining head loss calculations in pipes. For making this determination in open channels, the *Manning Equation* was developed during the latter part of the nineteenth century. Later, this equation was used for both open channels and closed conduits.

In the early 1900s, a more practical equation, the *Hazen-Williams* equation, was developed for use in making calculations related to water pipes and wastewater force mains:

$$Q = 0.435 \times CD^{2.63} \times S^{0.54}, \qquad (4.18)$$

where
Q = flow rate, (ft^3/s)
C = coefficient of roughness, (C decreases with roughness)
D = hydraulic radius R, (ft) and
S = slope of energy grade line (ft/ft).

C Factor

C factor, as used in the Hazen-Williams formula, designates the coefficient of roughness. *C* does not vary appreciably with velocity, and by comparing pipe types and ages, it includes only the concept of roughness, ignoring fluid viscosity and the Reynold's Number. Based on experience (experimentation), accepted tables of *C* factors have been established for pipe (see Table 4.4). Generally, *C* factor decreases by one with each year of pipe age. Flow for a newly designed system is often calculated with a *C* factor of 100, based on averaging it over the life of the pipe system.

DID YOU KNOW?

A high *C* factor means a smooth pipe. A low *C* factor means a rough pipe.

DID YOU KNOW?

An alternate to calculating the Hazen-Williams formula, called an alignment chart, has become quite popular for fieldwork. The alignment chart can be used with reasonable accuracy.

Slope

Slope is defined as the head loss per foot. In open channels, where the water flows by gravity, slope is the amount of incline of the pipe, and is calculated as feet of drop per foot of pipe length (ft/ft). Slope is designed to be just enough to overcome frictional losses so that the velocity remains constant, the water keeps flowing, and solids will not settle in the conduit. In piped systems, where pressure loss for every foot of pipe is experienced, slope is not provided by slanting the pipe but instead by pressure added to overcome friction.

MINOR HEAD LOSS

In addition to the head loss caused by friction between the fluid and the pipe wall, losses also are caused by turbulence created by obstructions (i.e., valves and fittings of all types) in the line, changes in direction, and changes in flow area.

DID YOU KNOW?

In practice, if minor head loss is less than 5% of the total head loss, it is usually ignored.

TABLE 4.4
C Factors

Type of Pipe	C Factor
Asbestos Cement	140
Brass	140
Brick Sewer	100
Cast Iron	110
10 years old	90
20 years old	
Ductile Iron, (cement lined)	140
Concrete or Concrete Lined	140
Smooth, steel forms	120
Wooden forms	110
Rough	
Copper	140
Fire Hose (rubber lined)	135
Galvanized Iron	120
Glass	140
Lead	130
Masonry Conduit	130
Plastic	150
Steel	150
Coal-tar enamel lined	140
New Unlined	110
Riveted	
Tin	130
Vitrified	120
Wood Stave	120

Source: Lindeburg (1986).

RESERVOIR STORED ENERGY

One of the major components of a hydroelectric dam is the area behind the dam, its reservoir. The water temporarily stored there is called gravitational potential energy. The water is stored in a position above the rest of the dam facility to allow gravity to carry the water down to the turbines. Because this higher altitude is different than where the water would naturally be, the water is at an altered equilibrium. The result is stored energy of position, that is, gravitational potential energy. The water has the potential to do work because of the position it is in. As shown in Figure 4.3, gravity will force the water to fall to a lower position through the intake and the control gate. Installed within the dam, when the control gate is opened, the water from the reservoir goes through the intake and becomes translational kinetic energy as it falls through the next main part of the system, the penstock. Translational kinetic energy is the energy

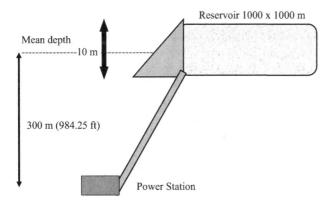

Reservoir 1000 x 1000 m

Mean depth

----------------10 m

300 m (984.25 ft)

Power Station

FIGURE 4.3 Schematic representation of a hydro scheme. Adapted from Tovey (2005).

due to motion from one location to another. The water is moving (falling) from the reservoir toward the turbines through the penstock; the water is carried through the penstock's long shaft toward the turbines where the kinetic energy becomes mechanical energy. The force of the water is used to turn the turbines that turn the generator shaft. The generators convert the energy of water into electricity and then step-up transformers increase the voltage produced to higher voltage levels.

As mentioned, potential energy stored in the reservoir is converted into kinetic energy at the inlet to the water turbine. Thus we can equate:

$$mgh = \tfrac{1}{2} \, mV^2, \tag{4.19}$$

where

m = mass of water,
g = acceleration due to gravity (can be taken as 10mm^{-2}) in most applications,
h = head difference, and
V = velocity of water at the inlet.

EXAMPLE 4.13[3]

Problem:

A reservoir has an area of 1 sq km, and the difference between the crest of the dam and the inlet to a hydro station is 10m (see Figure 4.3). The station runs at an overall efficiency of 80% and is situated 305m below the crest of the dam. The rainfall is 1000mm per annum, the catchment area of the reservoir is 10 times the area of the reservoir, and the run is 50%. What should the rated output of the turbine be if its maximum output is designed to be 5 times the mean output at the site? What is the maximum time the station could operate at full power during a sustained drought?

Solution:

Mean head between max and min levels = 305 — 10/2 = 300m
Average annual flow into reservoir

> = 50% of 10 times area multiplied by rainfall
> = 0.5 x 10 x 1000 x 1000 x 1 = 5,000,000 m³.

Mean energy generated per annum at 80% efficiency = [mgh x 0.8]

> = 5,000,000 x 1000 x 10 x 300 x 0.8
> = 12,000,000 MJ.

Rated output (i.e., mean power)

> = 12,000,000/60 x 60 x 24 x 365 = 0.381 MW.

So max power out = 5 x 0.381 = 190 MW

and time at max power assuming reservoir falls by 10m

$$= \frac{\text{area} \times \text{depth} \times \text{density} \times gh \times 0.8}{\text{max power}}$$

$$= \frac{1000 \times 1000 \times 10 \times 1000 \times 300 \times 0.8}{1900000 \times 60 \times 60 \times 24}$$

= 146.2 days.

HYDRO TURBINES

According to EERE (2008), there are two main types of hydro turbines: impulse and reaction. The type of hydropower turbine selected for a project is based on the height of standing water—as mentioned, this is referred to as "head"—and the flow, or volume of water, at the site. Cost, efficiency desired, and how deep the turbine must be set are other deciding factors.

Figure 4.4 shows the powerhouse at Rocky Reach Dam, Washington.

IMPULSE TURBINE

The impulse turbine uses the velocity of the water to move the runner and discharges atmospheric pressure. The water stream hits each bucket on the runner. There is no suction on the downside of the turbine, and water flows out the bottom of the turbine housed after hitting the runner. An impulse turbine is generally suitable for high-head, low-flow applications.

FIGURE 4.4   Powerhouse at Rocky Reach Dam, Columbia River, Washington. Photo by F.R. Spellman.

REACTION TURBINE

A reaction turbine develops power from the combined action of pressure and moving water. The runner is placed directly in the water stream flowing over the blades rather than striking each individually. Reaction turbines are generally used for sites with lower head and higher flows than compared with impulse turbines.

ADVANCED HYDROPOWER TECHNOLOGY

The USDOE and its associated technical activities support the development of technologies that will enable existing hydropower to generate more electricity with less environmental impact. This will be done by 1) developing new turbine systems that have improved overall performances, 2) developing new methods to optimize hydropower operations at the unit, plant, and reservoir system levels, and 3) conducting research to improve the effectiveness of the environmental mitigation practices required at hydropower projects.

The USDOE (2008) reports that the main objective of its research into advanced hydropower technology is to develop new system designs and operation modes that will enable both better environmental performance and competitive generation of electricity. The products of the USDOE's research will allow hydropower projects to generate cleaner electricity. USDOE-sponsored projects will develop and demonstrate new equipment and operational techniques that will optimize water-use efficiency, increase generation, and improve environmental performance and mitigation practices in existing plants. Ongoing research efforts contributing to the success of these objectives will enable up to a 10% increase in the hydropower generation at existing dams. These objectives include:

- Testing a new generation of large turbines in the field to demonstrate that these turbines are commercially viable, compatible with today's environmental standards, and capable of balancing environmental, technical, operational, and cost considerations.
- Developing new tools to improve water use efficiency and operations optimization within hydropower units, and plants and river systems with multiple hydropower facilities.
- Identifying improved practices that can be applied at hydropower plants to mitigate for environmental effects of hydro development and operation.

HYDROPOWER GENERATION: DISSOLVED OXYGEN CONCERNS

Regarding the benefits derived from the use of hydropower: it is a clean fuel source; it is a fuel source that is domestically supplied; it relies on the water cycle and thus is a renewable power source; it is generally available as needed; it creates reservoirs that offer a variety of recreational opportunities, notably fishing, swimming, and boating; and they supply water where needed and assist in flood control—many of these are well known and often taken for granted.

Coins are two-sided, of course, and so are the facts about hydropower. That is, with the good side of anything, there generally is an accompanying bad side. Many view this to be the case with hydropower. The bad side, or disadvantages of hydropower include the impact on fish populations (e.g., salmon) if they can't migrate upstream past impoundment dams to spawning grounds, or if they can't migrate downstream to the ocean. Hydropower can also be impacted by drought in that when water is not available, the plant can't produce electricity. Hydropower plants also compete with other uses for the land.

Other lesser-known negatives of hydropower plants concern their impact on water flow and quality; hydropower plants can cause low water levels that impact riparian habitats. Water quality is also affected by hydropower plants. The low dissolved oxygen levels in the water, a problem that is harmful to riparian (riverbank) habitats, can result when reservoirs stratify (develop layers of water of different temperatures). Stratification could affect the water temperature with resultant effects on dissolved oxygen levels, nutrient levels, productivity, and the bioavailability of heavy metals. During the summer, stratification, a natural process, can divide the reservoir into distinct vertical strata, that is, a warm, well-mixed upper layer (epilimnion) overlying a cooler, relativity stagnant lower layer (hypolimnion). Plant and animal respiration, bacterial decomposition of organic matter, and chemical oxidation can all act to progressively remove DO from hypolimnetic waters. This decrease in hypolimnetic DO is not generally offset by the renewal mechanisms of atmospheric diffusion, circulation, and photosynthesis that operate in the epilimnion (Spellman 1996; 2008). In temperate regions, the decline in hypolimnetic DO concentrations begins at the onset of stratification (spring or summer) and continues until either anaerobic conditions predominate or reoxygenation occurs during the fall turnover of the water body.

There are numerous structural, operational, and regulatory techniques that a hydropower operator can use to resolve a low DO issue. Levels of DO can be increased through modifications in dam operations. These include such techniques as fluctuating the timing and duration of flow releases; spilling or sluicing water; increasing minimum flows, flow mixing, and turbine aeration; and, at some sites, injecting air or oxygen in weir aeration, which has proven effective. The most effective strategy for addressing the DO problem is dependent on the site-specific situation.

BOTTOM LINE ON HYDROPOWER

Hydropower offers advantages over the other energy sources but faces unique environmental challenges. As mentioned, the advantages of using hydropower begin with the fact that hydropower does not pollute the air like power plants that burn fossil fuels, such as coal and natural gas. Moreover, hydropower does not have to be imported into the United States like foreign oil does; it is produced in the United States. Because hydropower relies on the water cycle, driven by the sun, it's a renewable resource that will be around for at least as long as humans. Hydropower is controllable. That is, engineers can control the flow of water through the turbines to produce electricity on demand. Finally, hydropower impoundment dams create huge lake areas for recreation, irrigation of farmlands, reliable supplies of potable water, and flood control.

Again, as mentioned, hydropower also has some disadvantages. For example, fish populations can be impacted if fish cannot migrate upstream past impoundment dams to spawning grounds, or if they cannot migrate downstream to the ocean. Many dams have installed fish ladders or elevators to aid upstream fish passage. Downstream fish passage is aided by diverting fish from turbine intakes using screens or racks or even underwater lights and sounds, and by maintaining a minimum spill flow past the turbine. Hydropower can also impact water quality and flow. Hydropower plants can cause low dissolved oxygen (DO) levels in the water, a problem that is harmful to riparian (riverbank) habitats and is addressed using various aeration techniques, which oxygenates the water. Maintaining minimum flows of water downstream of a hydropower installation is also critical for the survival of riparian habitats. Hydropower is also susceptible to drought. When water is not available, the hydropower plants can't produce electricity. Finally, construction of new hydropower facilities impacts investors and others by competing with other uses of the land. Preserving local flora and fauna and historical or cultural sites is often more highly valued than electricity generation.

NOTES

1   Much of the information in this section is from F. R. Spellman (2008). *The Science of Water*. Boca Raton, FL: CRC Press.
2   From J. A. Nathanson, *Basic Environmental Technology: Water Supply, Waste Management, and Pollution Control*, 2nd ed. Upper Saddle River, NJ: Prentice Hall, pp. 21–22, 1997.

3 Example modified from Tovey, 2005. *ENV-2E02 Energy Resources—Lecture Notes*, p. 109. Accessed 03/01//19 @ www2.env.ac.uk.gmmc/energy.

REFERENCES AND RECOMMENDED READING

Efficiency & Renewable Energy (EERE). 2008. *History of Hydropower*. Accessed 06/15/19 @ www.eere.energy.gov/windandhydro/printable_versions/hydro_hisotry.html
Gibilisco, S. 2007. *Alternative Energy Demystified*. New York: McGraw-Hill.
Lindeburg, M.R. 1986. *Civil Engineering Reference Manual*, 16th edition. Los Angeles: PPI.
Spellman, F.R. 1996. *Stream Ecology & Self-Purification*. Lancaster, PA: Technomic Publishing Company.
Spellman, F.R. 2008. *The Science of Water*, 2nd edition. Boca Raton, FL: CRC Press.
Tovey, N.K., 2005. *ENV-2E02 Energy Resources 2005 Lecture*. Accessed 03/01/10 @ www2.env.ac.UK.gmmc/energy
US Department of Energy (USDOE). 2008. *Advanced Hydropower Technology*. Accessed 03/01/19 @ wwwl.eere.energy.gov/windlandhydro/printable_versions/hydro_advtech.htm

5 Bioenergy

A Nation that runs on oil can't afford to run short.

—Old Oil Industry Slogan

INTRODUCTION

In a *New York Times* article, Thomas Friedman (2010) stated that "the fat lady has sung." Specifically, Friedman was speaking about America's transition from "The Greatest Generation" to what Kurt Anderson referred to as "The Grasshopper Generation." That is, according to Friedman, we are "going from the age of government handouts to the age of citizen givebacks, from the age of companions fly free to the age of paying for each bag." He goes on to say that we all accept that our parents were the greatest generation, but it is us that we are concerned about and that it is "we" who comprise the Grasshopper Generation: "we are eating through the prosperity that was bequeathed us like hungry locusts."

As mentioned (and the major theme of this text), what we are "eating through," among other things, is our readily available, relatively cheap source of energy. The point is we can, like the grasshopper, gobble it all up until it is all gone, or we can find alternatives—renewable alternatives of energy.

One of the promising sources of energy is bioenergy. Bioenergy is a general term for energy derived from materials such as straw, wood, or animal wastes, which, in contrast to fossil fuels, were living matter relatively recently. Such materials can be burned directly as solids (biomass) to produce heat or power but can also be converted into liquid biofuels. In the last few years there has been much increased interest in bioenergy fuels such as biofuels that can be used for transport. That is, biofuels, as mentioned, are a liquid fuel (biodiesel and bioethanol) used for transport. Now, it is transport that takes center stage in finding renewable, alternate fuels to eventually replace hydrocarbon fuels. Unlike biofuels, solid biomass fuel is used primarily for electricity generation and/or heat supply.

Even though we have stated that bioenergy is one of the promising sources of energy in and for the future (and it is), it is rather ironic whenever the experts (or anyone else for that matter) state this point without qualification. The qualification?

DOI: 10.1201/9781003439059-6

The reality? Simply, keep in mind that it was only 100 years ago that our economy was based primarily on bioenergy from biomass, or carbohydrates, rather than from hydrocarbons. The USDOE (2003a) points out that in the late 1800s, the largest selling chemicals were alcohols made from wood and grain; the first plastics were produced from cotton; and about 65% of the nation's energy came from wood.

By the 1920s, the economy started shifting toward the use of fossil resources, and after World War II this trend accelerated as technological breakthroughs were made. By the 1970s, fossil energy was established as the backbone of the US economy, and all but a small portion of the carbohydrate economy remained. In the industrial sector, plants accounted for about 16% of input in 1989, compared with 35% in 1925.

Processing cost and the availability of inexpensive fossil energy resources continue to be driving factors in the dominance of hydrocarbon resources. In many cases, it is still more economical to produce goods from petroleum or natural gas than from plant matter. This trend is about to shift dramatically as we reach peak oil and as the world continues to demand unprecedented amounts of petroleum supplies from an ever-dwindling supply.

Assisting in this trend shift are the technological advances in the biological sciences and engineering, political change, and concern for the environment that have begun to swing the economy back toward carbohydrates on several fronts. Consumption of biofuels in vehicles, for example, has risen from zero in 1977 to nearly 1.5 billion gallons in 1999. The use of inks produced from soybeans in the United States increased by fourfold between 1989 and 2000 and is now at more than 22% of total use (ILSR, 2002).

Technological advances are also beginning to make an impact on reducing the cost of producing industrial products and fuels from biomass, making them more competitive with those produced from petroleum-based hydrocarbons. Developments in pyrolysis, ultra-centrifuges, membranes, and the use of enzymes and microbes as biological factories are enabling the extraction of valuable components from plants at a much lower cost. As a result, industry is investing in the development of new bioproducts that are steadily gaining a share of current markets (USDOE 2003a).

New technology is enabling the chemical and food processing industries to develop new processes that will enable more cost-effective production of all kinds of industrial products from biomass. One example is a plastic polymer derived from corn that is now being produced at a 300-million-pound-per-year plant in Nebraska, a joint venture between Cargill, the largest grain merchant, and Dow Chemical, the largest chemical producer (Fahey, 2001).

Other chemical companies are exploring the use of low-cost biomass processes to make chemicals and plastics that are now made from more expensive petrochemical processes (USDOE, 2003b). In this regard, new innovative processes such as biorefineries may become the foundation of the new bioindustry. The biorefinery is similar in concept to the petroleum refinery, except that it is based on conversion of biomass feedstocks rather than crude oil. Biorefineries in theory would use multiple forms of biomass to produce a flexible mix of products, including fuels, power, heat, chemicals, and materials (see Figure 5.1). In a biorefinery, biomass would be converted into high-value chemical products and fuels (both gas and liquid). By-products and

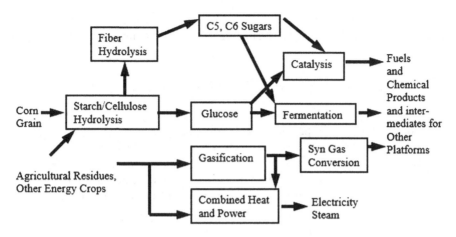

FIGURE 5.1 The biorefinery.

TABLE 5.1

US Energy Consumption by Energy Source, 2022 (%)

Energy Source	2022
Total	**97.33**
Renewable	12.16
Biomass (biofuels, waste, wood and wood-derived)	**40%**
Biofuels	**19%**
Waste	**4%**
Wood-derived Fuels	**17%**
Geothermal	2%
Hydroelectric Conventional	19%
Solar/PV	12%
Wind	27%

Source: EIA 2022. *U. Energy consumption by Energy Source.* Accessed 06/25/23 @http://www.eia.ener
gyexplained/us-energy-facts.

---

residues, as well as some portions of the fuels produced, would be used to fuel on-site power generation or cogeneration facilities producing heat and power.

The biorefinery concept has already proven successful in the US agricultural and forest products industries, where such facilities now produce food, feed fiber, or chemicals, as well as heat and electricity to run plant operations.

Biorefineries offer the most potential for realizing the ultimate opportunities of the bioenergy industry.

Table 5.1 highlights bioenergy's quadrillion Btu ranking in current renewable energy source use. As pointed out in the table, the energy consumption by energy source computations are from the year 2022. Thus, the biomass (biofuels, wastes, wood and wood-derived) 40% is expected to steadily increase to an increasingly higher level and this should be reflected in the 2022–2023 figures when they are released.

DID YOU KNOW?

In 2004, Hanson estimated that in the United States alone more than 100 billion gallons of light fuel oil per year can be produced from waste and biomass. The source waste and biomass included in this estimation included municipal solid waste, municipal sewage sludge, Hazmat waste, agricultural crop waste, feedlot manure, plastics, tires, heavy oil or tar sands, forestry waste, restaurant grease and biomass crops (switchgrass) growing on idle land and cropland.

BIOMASS

Biomass (all Earth's living matter) consists of the energy from plants and plant-derived organic-based materials; it is essentially stored energy from the sun. Biomass can be biochemically processed to extract sugars, thermochemically processed to produce biofuels or biomaterial, or combusted to produce heat to electricity. Biomass is also an input into other end-of-use markets, such as forestry products (pulpwood) and other industrial applications. This complicates the economics of biomass feedstock and requires that we differentiate between what is technically possible from what is economically feasible, considering relative prices and intermarket competition.

Biomass has been used since people began burning wood to cook food and keep warm. Trees have been the principal fuel of almost every society for over 5,000 years, from the Bronze Age until the middle of the nineteenth century. Wood is still the largest biomass energy resource today, but other sources of biomass can also be used. These include food crops, grassy and woody plants, residues from agriculture or forestry, and the organic components of municipal and industrial wastes. Even the fumes from landfills (which are methane, a natural gas) can be used as a biomass energy source. It excludes organic material that has been transformed by geological processes into substances such as coal or petroleum. The biomass industry is one of the fastest-growing industries in the United States.

FEEDSTOCK TYPES

A variety of biomass feedstocks can be used to produce transportation fuels, biobased products, and power. Feedstocks refer to the crops or products, like waste vegetable oil, that can be used as or converted into biofuels and bioenergy. Regarding the advantages or disadvantages of one type of feedstock as compared to another, this is gauged in terms of how much usable material they yield, where they can grow, and how energy and water-intensive they are. Feedstock types are listed as first-generation or second-generation feedstocks. First-generation feedstocks include those that are already widely grown and used for some form of bioenergy or biofuel production, which means that there are possible foods versus fuel conflicts. First-generation feedstocks include sugars (sugar beets, sugar cane, sugar palm, sweet sorghum, and nypa palm), starches (cassava, corn, milo, sorghum, sweet potato, and wheat), waste feedstocks such as whey and citrus peels, and oils and fats (coconut oil, oil palm, rapeseed, soybeans sunflower seed, castor beans jatropha, jojoba, karanj, waste vegetable oil and animal fat). Second-generation feedstocks refers broadly to crops that have high potential yields of biofuels, but that are not widely cultivated, or not cultivated as an energy crop. It refers to cellulosic feedstocks or conventional crops such as miscanthus, prairie grass, and switchgrass, willow and hybrid popular trees. Algae and halophytes (saltwater plants) are other second-generation feedstocks.

Currently, most of the ethanol produced in the United States is made from corn or other starch-based crops. The present focus, however, is on the development of cellulosic feedstocks—non-grain, non-food-based feedstocks such as switchgrass, corn stover, and wood material—and on technologies to convert cellulosic material into transportation fuels and other products. Using cellulosic feedstocks can not only alleviate the potential concern of diverting food crops to produce fuel, but also has a variety of environmental benefits (EERE, 2008).

Because such a wide variety of cellulosic feedstocks can be used for energy production, potential feedstocks are grouped into categories, or pathways. Figure 5.2 shows some of the specific feedstocks in each of these areas.

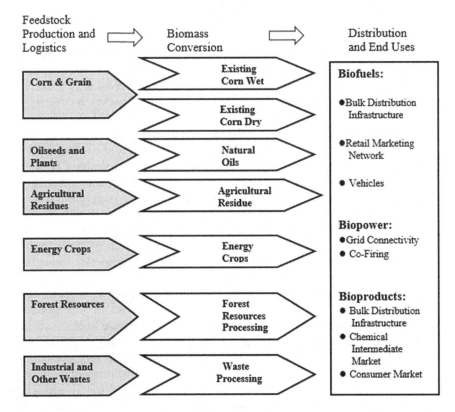

FIGURE 5.2   Resource-based biorefinery pathways.

Composition of Biomass

The ease with which biomass can be converted to useful products or intermediates is determined by the composition of the biomass feedstock. Biomass contains a variety of components, some of which are readily accessible and others that are much more difficult and costly to extract. The composition and subsequent conversion issues for current and potential biomass feedstock compounds are listed and described below.

- Starch (Glucose)—is readily recovered and converted from grain (corn, wheat, rice) into products. Starch from corn grain provides the primary feedstock for today's existing and emerging sugar-based bioproducts, such as polylactide as well as the entire fuel ethanol industry. Corn grain serves as the primary feedstock for starch used to manufacture today's biobased products. Core wet mills use a multistep process to separate starch from the germ, gluten (protein), and fiber components of corn grain. The starch streams generated by wet milling are highly pure, and acid of enzymatic hydrolysis is used to break the glycosidic linkages of starch to yield glucose. Glucose is then converted into a multitude of useful products.
- Lignocellulosic Biomass—the non-grain portion of biomass (e.g., cobs, stalks), often referred to as agricultural stover or resides, and energy crops such as switchgrass also contain valuable components, but they are not as readily accessible as starch. These lignocellulosic biomass resources (also called cellulosic) are comprised of cellulose, hemicellulose, and lignin. Generally, lignocellulosic material contains 30%–50% cellulose, 20%–30% hemicellulose, and 20%–30% lignin. Some exceptions to this are cotton (98% cellulose) and flax (80% cellulose). Lignocellulosic biomass is perceived as a valuable and largely untapped resource for the future bioindustry. However, recovering the components in a cost-effective way represents a significant technical challenge.
- Cellulose—is one of nature's polymers and is composed of glucose, a six-carbon sugar. The glucose molecules are joined by glycosidic linkages, which allow the glucose chains to assume an extended ribbon conformation. Hydrogen bonding between chains leads to the formation of the flat sheets that lay on top of one another in a staggered fashion, like the way staggered bricks add strength and stability to a wall. As a result, cellulose is very chemically stable and insoluble and serves as a structural component in plant walls.
- Hemicellulose—is a polymer containing primarily 5-carbon sugars such as xylose and arabinose with some glucose and mannose dispersed throughout. It forms a short-chain polymer that interacts with cellulose and lignin to form a matrix in the plant wall, strengthening it. Hemicellulose is more easily hydrolyzed than cellulose. Much of the hemicellulose in lignocellulosic material is solubilized and hydrolyzed to pentose and hexose sugars
- Lignin—helps bind the cellulosic/hemicellulose matrix while adding flexibility to the mix. The molecular structure of lignin polymers is very random and disorganized and consists primarily of carbon ring structures (benzene rings with methoxyl, hydroxyl, and propyl groups) interconnected by polysaccharides (sugar polymers). The ring structures of lignin have great potential as valuable

chemical intermediates. However, separation and recovery of the lignin is difficult.

- Oils and Protein—the seeds of certain plants offer two families of compounds with great potential for bioproducts: oils and protein. Oils and protein are found in the seeds of certain plants (soybeans, castor beans), and can be extracted in a variety of ways. Plants raised for this purpose include soy, corn, sunflower, safflower, rapeseed, and others. A large portion of the oil and protein recovered from oilseeds and corn is processed for human or animal consumption, but they can also serve as raw materials for lubricants, hydraulic fluids, polymers, and a host of other products.
- Vegetable Oils—are composed primarily of triglycerides, also referred to as triacylglycerols. Triglycerides contain a glycerol molecule as the backbone with three fatty acids attached to glycerol's hydroxyl groups.
- Proteins—are natural polymers with amino acids as the monomer unit. They are incredibly complex materials and their functional properties depend on molecular structure. There are 20 amino acids each differentiated by their side chain or R-group and they can be classified as nonpolar and hydrophobic, polar uncharged, and ionizable. The interactions among the side chains, the amide protons, and the carbonyl oxygen help create the protein's 3D shape.

UNDERSTANDING BIOMASS: PLANT BASICS[1]

Optimizing plant biomass for more efficient processing requires a better understanding of plants and plant cell-wall structure and function. The plant kingdom ranks second in importance only to the animal kingdom (at least from the human point of view). The importance of plants and plant communities to humans, bioenergy production, and their environment cannot be overstated. Some of the important things plants provide are listed below:

- *Aesthetics*—plants add to the beauty of the places we live.
- *Medicine*—80% of all medicinal drugs originate in wild plants.
- *Food*—90% of the world's food comes from only 20 plant species.
- *Industrial Products*—plants are very important for the goods they provide (e.g., plant fibers provide clothing), and wood is used to build homes.
- *Recreation*—plants form the basis for many important recreational activities, including fishing, nature observation, hiking, and hunting.
- *Air Quality*—the oxygen in the air we breathe comes from the photosynthesis of plants.
- *Water Quality*—plants aid in maintaining healthy watersheds, streams, and lakes by holding soil in place, controlling stream flows, and filtering sediments from water.
- *Erosion Control*—plant cover helps prevent wind or water erosion of the top layer of soil that we depend on.
- *Climate*—regional climates are impacted by the amount and type of plant cover.

- *Fish and Wildlife Habitat*—plants provide the necessary habitat for wildlife and fish populations.
- *Ecosystem*—every plant species serves an important role or purpose in their community.
- *Feedstock for bioenergy production*—some important fuel chemicals come from plants, such as ethanol from corn and soy diesel from soybeans.

Though both are important kingdoms of living things, plants and animals differ in many important aspects. Some of these differences are summarized in chart below.

Plants	Animals
Plants contain chlorophyll and can make their own food.	Animals cannot make their own food and are dependent on plants and other animals for food.
Plants give off oxygen and take in carbon dioxide given off by animals.	Animals give off carbon dioxide, which plants need to make food, and take in oxygen, which they need to breathe.
Plants generally are rooted in one place and do not move on their own.	Most animals can move fairly freely.
Plants have either no or very basic ability to sense.	Animals have a much more highly developed sensory and nervous system.

Before discussing the basic specifics of plants, it is important to first define a few key plant terms.

Plant Terminology

- **Apical meristematic cells**—consists of meristematic cells located at tip (apex) of a root or shoot.
- **Cambium**—the lateral meristem in plants.
- **Chloroplasts**—disk-like organelles with a double membrane found in eukaryotic plant cells.
- **Companion cells**—specialized cells in the phloem that load sugars into the sieve elements.
- **Cotyledons**—leaf-like structure (sometimes referred to as seed leaf) that is present in the seeds of flowering plants.
- **Dicot**—one of the two main types of flowering plants; characterized by having two cotyledons.
- **Diploid**—having two of each kind of chromosome (2n).
- **Guard cells**—specialized epidermal cells that flank stomata and whose opening and closing regulate gas exchange and water loss.
- **Haploid**—having only a single set of chromosomes (n).
- **Meristem**—a group of plant cells that can divide indefinitely provides new cells for the plant.

TABLE 5.2
The Main Phyla/Division of Plants

Phylum/Division	Examples
Bryophyta	mosses, liverworts and hornworts
Coniferophyta	conifers such as redwoods, pines and firs
Cycadophyta	cycads, sago palms
Gnetophyta	shrub trees and vines
Ginkophyta	*Ginkgo* is the only genus
Lycophyta	lycopods (look like mosses)
Pterophyta	ferns and tree-ferns
Anthophyta	flowering plants including oak, corn, maize and herbs

- **Monocots**—one of two main types of flowering plants; characterized by having a single cotyledon.
- **Periderm**—a layer of plant tissue derived from the cork cambium, and then secondary tissue, replacing the epidermis.
- **Phloem**—complex vascular tissue that transports carbohydrates throughout the plant.
- **Sieve cells**—conducting cells in the phloem of vascular plants.
- **Stomata**—pores on the underside of leaves that can be opened or closed to control gas exchange and water loss.
- **Thallus**—main plant body, not differentiated into a stem or leaves.
- **Tropism**—plant behavior; controlling the direction of plant growth.
- **Vascular tissue**—tissues found in the bodies of vascular plants that transport water, nutrients, and carbohydrates. The two major kinds are xylem and phloem.
- **Xylem**—vascular tissue of plants that transports water and dissolved minerals from the roots upward to other parts of the plant. Xylem often also provides mechanical support against gravity.

Although not typically acknowledged, plants are as intricate and complicated as animals. Plants evolved from photosynthetic protists and are characterized by photosynthetic nutrition, cell walls made from cellulose and other polysaccharides, lack of mobility, and a characteristic life cycle involving an alternation of generations. The phyla/division of plants and examples are listed in Table 5.2.

THE PLANT CELL

A summary of eukaryote plant cells (have a membrane-bound nucleus and organelles) is provided here (see Figure 5.3).

- *Plants have all the organelles animal cells have* (i.e., nucleus, ribosomes, mitochondria, endoplasmic reticulum, Golgi apparatus, etc.).

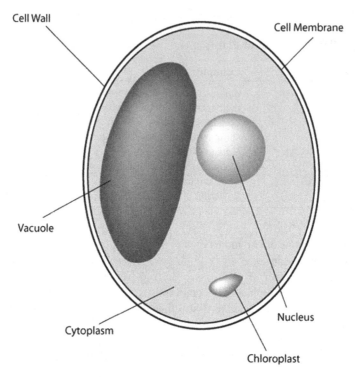

Cell Wall

Cell Membrane

Vacuole

Cytoplasm

Nucleus

Chloroplast

FIGURE 5.3   Eukaryote: plant cell.

- *Plants have chloroplasts.* Chloroplasts are special organelles that contain chlorophyll and allow plants to carry out photosynthesis.
- *Plant cells can sometimes have large vacuoles for storage.*
- *Plant cells are surrounded by a rigid cell wall made of cellulose,* in addition to the cell membrane that surrounds animal cells. Those walls provide support.

Vascular Plants

Vascular plants, also called **Tracheophytes**, have special vascular tissue for transport of necessary liquids and minerals over long distances. Vascular tissues are composed of specialized cells that create "tubes" through which materials can flow throughout the plant body. These vessels are continuous throughout the plant, allowing for the efficient and controlled distribution of water and nutrients. In addition to this transport function, vascular tissues also support the plant. The two types of vascular tissue are xylem and phloem.

- **Xylem** consists of a tube or a tunnel (pipeline) in which water and minerals are transported throughout the plant to leaves for photosynthesis. In addition to distributing nutrients, xylem (wood) provides structural support. After a time, the xylem at the center of older trees ceases to function in transport and takes on a supportive role only.

- **Phloem** tissue consists of cells called *sieve tubes* and *companion cells.* Phloem tissue moves dissolved sugars (carbohydrates), amino acids and other producers of photosynthesis from the leaves to other regions of the plant.

The two most important Tracheophytes are gymnosperms (*gymno* = naked; *sperma* = seed) and angiosperms (*angio* = vessel, receptacle, container).

- **Gymnosperms**—the plants we recognize as gymnosperms represent the sporophyte generation (i.e., the spore-producing phase in the life cycle of a plant that exhibits alternation of generation). Gymnosperms were the first tracheophytes to use seeds for reproduction. The seeds develop in protective structures called cones. A gymnosperm contains some cones that are female and some that are male. Female cones produce spores that, after fertilization, become eggs enclosed in seeds that fall to the ground. Male cones produce pollen, which is taken by the wind and fertilizes female eggs by that means. Unlike flowering plants, the gymnosperm does not form true flowers or fruits. Coniferous tress such as firs and pines are good examples of gymnosperms.
- **Angiosperms**—the flowering plants, are the most highly evolved plants and the most dominant in present times. They have stems, roots, and leaves. Unlike gymnosperms such as conifers and cycads, angiosperms' seeds are found in flowers. Angiosperm eggs are fertilized and develop into a seed in an ovary that is usually in a flower.

There are two types of angiosperms: monocots and dicots.

- **Monocots**—these angiosperms start with one seed-leaf (cotyledon); thus, their name, which is derived from the presence of a single cotyledon during embryonic development. Monocots include grasses, grains, and other narrow-leaved angiosperms. The main veins of their leaves are usually parallel and unbranched, the flower parts occur in multiples of three, and a fibrous root system is present. Monocots include orchids, lilies, irises, palms, grasses, and wheat, corn, and oats.
- **Dicots**—angiosperms in this group grow two seed-leaves (two cotyledons). Most plants are dicots, and include maples, oaks, elms, sunflowers, and roses. Their leaves usually have a single main vein or three or more branched veins that spread out from the base of the leaf.

LEAVES

The principal function of leaves is to absorb sunlight for the manufacturing of plant sugars in photosynthesis. The leaves' broad, flattened surfaces gather energy from sunlight, while apertures on their undersides bring in carbon dioxide and release oxygen. Leaves develop as a flattened surface to present a large area for efficient absorption of light energy. On its two exteriors, the leaf has layers of epidermal cells that secrete a waxy, nearly impermeable cuticle (chitin) to protect against water loss (dehydration) and fungal or bacterial attack. Gases diffuse in or out of the leaf through **stomata**, small openings on the underside of the leaf. The opening or closing of the stomata occurs through the swelling or relaxing of **guard cells**. If the plant wants to limit the diffusion of gases and the transpiration of water, the guard cells swell together and close the stomata. Leaf thickness is kept to a minimum so that gases that enter the leaf can diffuse easily throughout the leaf cells.

CHLOROPHYLL/CHLOROPLAST

The green pigment in leaves is **chlorophyll**. Chlorophyll absorbs red and blue light from the sunlight that falls on leaves. Therefore, the light reflected by the leaves is diminished in red and blue and appears green. The molecules of chlorophyll are large. They are not soluble in the aqueous solution that fills plant cells. Instead, they are attached to the membranes of disc-like structures, called **chloroplasts**, inside the cells. Chloroplasts are the site of photosynthesis, the process in which light energy is converted to chemical energy. In chloroplasts, the light absorbed by chlorophyll supplies the energy used by plants to transform carbon dioxide and water into oxygen and carbohydrates.

Chlorophyll is not a very stable compound; bright sunlight causes it to decompose. To maintain the amount of chlorophyll in their leaves, plants continuously synthesize it. The synthesis of chlorophyll in plants requires sunlight and warm temperatures. Therefore, during summer chlorophyll is continuously broken down and regenerated in the leaves of trees.

PHOTOSYNTHESIS

Because our quality of life, and indeed our very existence, depends on photosynthesis, it is essential to understand it. In photosynthesis, plants (and other photosynthetic autotrophs) use the energy from sunlight to create the carbohydrates necessary for cell respiration. More specifically, plants take water, carbon dioxide, and transform them into glucose and oxygen:

$$6CO_2 + 6H_2O + \text{light energy} \rightarrow C_6H_{12}O_6 + 6O_2.$$

This general equation of photosynthesis represents the combined effects of two different stages. The first stage is called the light reaction and the second stage is called the dark reaction. The **light reaction** is the photosynthesis process in which solar energy is harvested and transferred into the chemical bonds of ATP; it can only occur in light. The **dark reaction** is the process in which food (sugar) molecules are formed from carbon dioxide from the atmosphere with the use of ATP; it can occur in dark if ATP is present.

DID YOU KNOW?

Charles Darwin was the first to discuss how plants respond to light. He found that the new shoots of grasses bend toward the light because the cells on the dark side grow faster than the lighted side.

ROOTS

Roots absorb nutrients and water, anchor the plant in the soil, provide support for the stem, and store food. They are usually below ground and lack nodes, shoots, and leaves. There are two major types of root systems in plants. Taproot systems have a stout main root with a limited number of side-branching roots. Examples of taproot system plants are nut trees, carrots, radishes, parsnips, and dandelions. Taproots make transplanting difficult. The second type of root system, fibrous, has many branched roots. Examples of fibrous root plants are mostly grasses, marigolds, and beans. Radiating from the roots is a system of root hairs, which vastly increase the absorptive surface area of the roots. Roots also anchor the plant in the soil.

GROWTH IN VASCULAR PLANTS

Vascular plants undergo two kinds of growth (growth is primarily restricted to meristems), primary and secondary growth. **Primary growth** occurs relatively close to the tips of roots and stems. It is initiated by apical meristems and it is primarily involved in the extension of the plant body. The tissues that arise during primary growth are called primary tissues, and the plant body composed of these tissues is called the primary plant body. Most primitive vascular plants are entirely made up of primary tissues. **Secondary growth** occurs in some plants; secondary growth thickens the stems and roots. Secondary growth results from the activity of lateral meristems. Lateral meristems are called **cambia** (cambium) and there are two types:

1) **Vascular cambium**—gives rise to secondary vascular tissues (secondary xylem and phloem). The vascular cambium gives rise to xylem to the inside and phloem to the outside.
2) **Cork cambium**—which forms the **periderm** (bark). The periderm replaces the epidermis in woody plants.

PLANT HORMONES

Plant growth is controlled by plant hormones, which influence cell differentiation, elongation, and division. Some plant hormones also affect the timing of reproduction and germination.

- **Auxins**—affect cell elongation (tropism), apical dominance, and fruit drop or retention. Auxins are also responsible for root development, secondary growth

in the vascular cambium, inhibition of lateral branching, and fruit development. Auxin is involved in absorption of vital minerals and fall color. As a
leaf reaches its maximum growth, auxin production declines. In deciduous
plants, this triggers a series of metabolic steps, which causes the reabsorption
of valuable materials (such as chlorophyll) and their transport into the branch or
stem for storage during the winter months. Once chlorophyll is gone, the other
pigments typical of fall color become visible.

- **Kinins**—promote cell division and tissue growth in leaf, stem, and root. Kinins
  are also involved in the development of chloroplasts, fruits, and flowers. In addition, they have been shown to delay senescence (aging), especially in leaves,
  which is one reason that florists use cytokinins on freshly cut flowers—when
  treated with cytokinins they remain green, protein synthesis continues, and
  carbohydrates do not break down.
- **Gibberellins**—produced in the root-growing tips and acts as a messenger to
  stimulate growth, especially elongation of the stem, and can also end the dormancy period of seeds and buds by encouraging germination. Additionally,
  gibberellins play a role in root growth and differentiation.
- **Ethylene**—controls the ripening of fruits. Ethylene may insure that flows
  are carpellate (female), while gibberellin confers maleness on flowers. It also
  contributes to the senescence of plants by promoting leaf loss and other changes.
- **Inhibitors**—restrain growth and maintain the period of dormancy in seeds
  and buds.

TROPISMS: PLANT BEHAVIOR

Tropism is the movement (and growth in plants) of an organism in response to an
external stimulus. For example, tropisms, controlled by hormones, are a unique characteristic of sessile organisms such as plants that enable them to adapt to different
features of their environment—gravity, light, water, and touch—so that they can
flourish. There are three main tropisms:

- **Phototropism**—the tendency of plants growing or bending (moving) in
  response to light. Phototropism results from the rapid elongation of cells on the
  dark side of the plant, which causes the plant to bend in the opposite direction.
  For example, the stems and leaves of a geranium plant growing on the window-
  sill always turn toward the light.
- **Gravitropism**—refers to a plant's tendency to grow toward or against gravity.
  A plant that displays positive gravitropism (plant roots) will grow downward,
  toward the center of the Earth. That is, gravity causes the roots of plants to
  grow down so that the plant is anchored in the ground and has enough water
  to grow and thrive. Plants that display negative gravitropism (plant stems) will
  grow upward, away from the Earth. Most plants are negatively gravitropic.
  Gravitropism is also controlled by auxin. In a horizontal root or stem, auxin
  is concentrated in the lower half, pulled by gravity. In a positively gravitropic
  plant, this auxin concentration will inhibit cell growth on the lower side, causing

the stem to bend downward. In a negatively gravitropic plant, this auxin concentration will inspire cell growth on that lower side, causing the stem to bend upward.

- **Thigmotropism**—some people notice that their houseplants respond to thigmotropism (i.e., growing or bending in response to touch), growing better when they touch them and pay attention to them. Touch causes parts of the plant to thicken or coil as they touch or are touched by environmental entities. For instance, tree trunks grow thicker when exposed to strong winds, and vines tend to grow straight until they encounter a substrate to wrap around.

PHOTOPERIODISM

Photoperiodism is the response of an organism (e.g., plants) to naturally occurring changes in light during a 24-hour period. The site of perception of photoperiod in plants is leaves. For instance, sunflowers are known for their photoperiodism, or their ability to open and close in response to the changing position of the sun throughout the day.

All flowering plants have been placed in one of three categories with respect to photoperiodism: short-day plants, long-day plants, and day-neutral plants.

- **Short-day plants**—flowering promoted by day lengths shorter than a certain critical day length—includes poinsettias, chrysanthemums, goldenrod, and asters
- **Long-day plants**—flowering promoted by day lengths longer than a certain critical day length—includes spinach, lettuce, and most grains
- **Day-neutral plants**—flowering response insensitive to day length—includes tomatoes, sunflowers, dandelions, rice, and corn

PLANT REPRODUCTION

Plants can reproduce both sexually and asexually. Each type of reproduction has its benefits and disadvantages. A comparison of sexual and asexual plant reproduction is provided in the following.

- Sexual Reproduction:
  - Sexual reproduction occurs when a sperm nucleus from the pollen grain fuses with egg cell from ovary of pistil (pistil defined: the female reproductive structures in flowers, consisting of the stigma, style, and ovary).
  - Each brings a complete set of genes and produces genetically unique organisms.
  - The resulting plant embryo develops inside the seed and grows when seed is germinated.
- Asexual reproduction:
  - Occurs when a vegetative part of a plant, root, stem, or leaf gives rise to a new offspring plant whose genetic content is identical to the "parent plant."

An example would be a plant reproducing by root suckers, shoots that come from the root system. The breadfruit tree is an example.

- Asexual reproduction is also called vegetative propagation. It is an important way for plant growers to get many identical plants from one very quickly.
- By sexual reproduction plants can spread and colonize an area quickly (e.g., crabgrass).

BIOMASS PLANT CELL WALLS

Figure 5.3 shows the basic organelles within a standard plant cell, including the cell wall. It should be pointed out, however, that plants can have two types of cell walls, primary and secondary (see Figure 5.4). Primary cell walls contain cellulose consisting of hydrogen-bonded chains of thousands of glucose molecules, in addition to hemicellulose and other materials all woven into a network. Certain types of cells, such as those in vascular tissues, develop secondary walls inside the primary wall after the cell has stopped growing. These cell-wall structures also contain lignin, which provides rigidity and resistance to compression. The area formed by two adjacent plant cells, the middle lamella, typically is enriched with pectin.

Cellulose in higher plants is organized into microfibrils, each measuring about 3 to 6 nm in diameter and containing up to 36 glucan chains having thousands of glucose residues. Like steel girders stabilizing a skyscraper's structure, the primary cell wall's mechanical strength is due mainly to the microfibril scaffold (i.e., crystalline cellulose core) (USDOE, 2007).

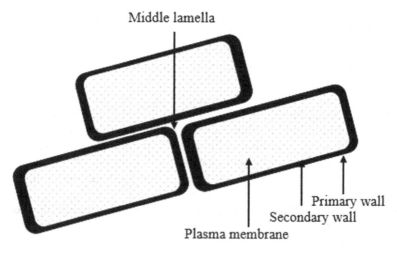

FIGURE 5.4   Plant cell walls.

DID YOU KNOW?

Cellulose microfibrils are composed of linear chains of glucose molecules that hydrogen binds to form the microfibrils.

FIRST-GENERATION FEEDSTOCKS[2]

The types of major end-use products are easily categorized for first-generation feedstocks (see Table 5.3). For ethanol production, corn and sugarcane are the most used feedstocks. Soybean and other vegetable oils and animal fats are used for biodiesel production (and bioproducts). Manure and landfill organic waste are used for methane production and the generator of electricity. Corn is used for ethanol production and currently is the leading feedstock used in the United States.

Several factors favor a positive outlook for further near-term growth in corn–ethanol production. Continued high oil prices will provide economic support for the expansion of all alternative fuel programs, including corn ethanol. Technology improvements that increase feedstock productivity and fuel conversion yields and positive spillovers from second-generation technologies (biomass gasification in ethanol refiners) will also help lower production costs for corn ethanol.

Among the factors likely to limit future growth of corn-ethanol production are increased feedstock and other production costs; increased competition from unconventional liquid fossil fuels (from oil sands, coal, heavy oil, and shale); the emergence of cellulosic ethanol as a low-cost competitor; and new policies to reduce greenhouse gas emissions (GHCs) that could favor advanced biofuels over corn ethanol.

Biodiesel is another biofuel experiencing expansion. While its production costs are higher than for ethanol, biodiesel has some environmental advantages, including biodegradability and lower sulfur and carbon dioxide emissions when burned. Biodiesel production in the United States increased rapidly from less than 2 million gallons in 2000 to about 500 million gallons in 2007. Policy incentives in the Energy Independence and Security Act of 2007 are expected to sustain demand for 1 billion gallons per year of this fuel after 2011.

A variety of oil-based feedstocks are converted to biodiesel using a process known as *transesterification*. This is the process of exchanging the organic group R" of an ester with the organic group R' of an alcohol. These reactions are often catalyzed by the addition of an acid or base as shown below:

Transesterification: alcohol + ester → different alcohol + different ester

The oil-based feedstocks include vegetable oils (mostly soy oil), recycled oils and yellow grease, and animal fats like beef tallow. It takes 3.4 kg of oil/fat to produce 1 gallon of biodiesel (Baise, 2006). Biodiesel production costs are high compared to ethanol, with feedstocks accounting for 80% or more of total costs.

TABLE 5.3
Biomass Feedstocks and Major Bioenergy End-use Applications

First-Generation Fuels		Second-Generation Fuels		
		Ethanol Biodiesel Methane	**Cellulosic ethanol**	**Thermoch Fuels (ethanol, diesel, butanol, etc.)**
First-generation				
Starch and sugar- feedstock for ethanol	Corn	X		
	Sugarcane	X		
	Molasses	X		
	Sorghum	X		
Vegetable oil and fats and biodiesel	Vegetable oils	X		
	Recycled fats & grease	X		
	Beef tallow	X		
Second-generation (short-term)				
Agricultural residues and livestock by- products	Corn stover		X	
	Wheat straw		X	
	Rice straw		X	
	Bagasse		X	
	Manure	X		
	Logging residues		X	X
Forest biomass	Fuel treatments		X	X
	Conventional wood		X	X
	Primary wood products		X	X
Urban woody waste & landfills	Secondary mill residues		X	X
	Municipal solid wastes	X	X	X
	Construction/demolition wood landfills	X	X	X
Second-generation (long-term)				
Herbaceous Energy Crops	Switchgrass		X	X
	Miscanthus		X	X
	Reed Canary Grass		X	X
	Sweet sorghum Alfalfa		X	X
	Willow		X	X
Short Rotation Woody Crops	Hybrid popular		X	X
	Cottonwood pines		X	X
	Sycamore pines		X	X
	Eucalyptus		X	X

The biodiesel industry consists of many small plants that are highly dispersed geographically. Decisions about plant location are primarily determined by local availability and access to the feedstock. Recent expansion in biodiesel production is affecting the soybean market. Achieving a nationwide target of a 2% biodiesel blend in diesel transportation fuel, for example, would require 2.8 million metric tons (MT) of vegetable oil, or about 30% of current US soybean oil production (USDA-World Agricultural Outlook Board, 2008).

One of the key biodiesel by-products is glycerin. At the present time there is concern about producing glycerin, mainly because there are no existing markets for the product. However, recent technological developments include an alternative chemical process that would produce biodiesel without glycerin. In addition, new processes are being tested that further transform glycerin into propylene glycol, which is used in the manufacture of antifreeze (Biodiesel, 2007).

SECOND-GENERATION FEEDSTOCKS: SHORT-TERM AVAILABILITY

Agricultural residues, a second-generation biomass feedstock, offer a potentially large and readily available biomass resource, but sustainability and conservation constraints could place much of it out of reach. Given current US cropland use, corn and wheat offer the most potentially recoverable residues. However, these residues play an important role in recycling nutrients in the soil and maintaining long-term fertility and land productivity. Removing too much residue could aggravate soil erosion and deplete the soil of essential nutrients and organic matter.

Safe removal rate methodologies, based on soil erosion, have been developed. Methodologies to determine removal rates while safeguarding soil fertility and meeting conservation objectives still need to be developed. Studies have shown that under current tillage practices, the national-average safe removal rate based on soil erosion for corn stover is less than 30%. Actual rates vary widely depending on local conditions. In other words, much of the generated crop residues may be out of reach for biomass use if soil conservation goals are to be achieved (Graham et al., 2007).

The estimated delivery costs for agricultural residues vary widely depending on crop type, load resource density, storage and handling requirements, and distance and transportation costs. Moreover, existing estimates are largely derived from engineering models, which may not account for economic conditions.

Agricultural residue feedstocks (such as corn stover) have a significant advantage in that they can be readily integrated into the expanding corn ethanol industry. However, dedicated energy crops (such as switchgrass) may have more benign environmental impacts.

Another significant biomass source is *forest biomass*, which can be immediately available should the bioenergy market develop. Logging residues are associated with timber industry activities and constitute significant biomass resources in many states, particularly in the Northeast, the north Central states, the Pacific Northwest, and the Southeast. In the Western States, the predominance of public lands and environmental pressure reduces the supply potential for logging residues, but there is a vast potential for biomass from thinning undertaken to reduce the risk of forest fires. However, the few analyses that have examined recoverability of logging residues cite the need to account for factors such as the scale and location of biorefineries and biopower plants, as well as regional resource density.

The potential for forest residues may be large but the actual quantities available for biomass conversion may be low due to the economics of harvesting, handling, and transporting the residues from forest areas to locations where they could be used. It is not clear how these residues compete with fossil fuels in the biopower and co-firing industries. In addition, there are competing uses for these products in the pulp and paper industry, as well as different bioenergy end uses. Economic studies of logging residues suggest a current lack of competitiveness with fossil fuels (coal, gas). But logging residues could become more cost competitive with further improvements in harvesting and transportation technologies and with policies that require a fuller accounting of the social and environmental benefits from converting forest residues to biopower or biofuels.

Another source of forest residues that could be recovered in significant quantities is biomass from fuel treatments and thinning. Fuel treatment residues are the by-product of efforts to reduce risk of loss from fire, insects, and disease; and therefore present substantially different challenges than logging residues.

The overall value of forest health benefits such as clean air and water is generally believed to exceed the cost of treatment. However, treated forests are often distant from end-use markets, resulting in high transportation costs to make use of the harvested material. Road or trail access, steep terrain, and other factors commonly limit thinning operations in Western forests.

Transportation costs can be a significant factor in the cost of recovering biomass. As much as half the cost of the material delivered to a manufacturing facility may be attributed to transportation. The offset to high-cost transportation of forest thinning is on-site densification of the biomass. This could entail pelletization, fast pyrolysis (to produce bio-oil), or baling.

The economics of transporting thinned woody residues versus on-site densification depend on the distance to end-use markets. Densification may be more economical if power generation facilities are far away. In addition to co-firing or co-generation facilities, improvements in thermochemical conversion efficiency and establishment of small-scale conversion facilities using gasification and/or pyrolysis may favor the use of forest residues for biofuel production (Polagye et al., 2007).

A third major category of immediately available second-generation biomass is wood residues from *secondary mill products* and *urban wood waste*. Urban wood waste provides a relatively cheap feedstock to supplement other biomass resources (Wiltsee, 1998). Urban wood waste encompasses the biomass portion of commercial, industrial, and municipal solid waste (MSW), while secondary mill residues include sawdust, shavings, wood trims, and other by-products generated from processing lumber, engineered wood products, or wood particles. Both urban wood waste and secondary mill residues have several primary uses and disposal methods. Urban wood waste not used in captive markets (such as the pulpwood industry) could be used as biomass either to generate electricity or to produce cellulosic ethanol when it becomes commercially viable.

The amount of urban wood wastes produced in the United States is significant and their use as biomass could be economically viable, particularly in large urban centers (Wiltsee, 1998). Several national availability estimates exist for various types of urban wood wastes, but estimates vary depending on methodology, product coverage, and assumptions about alternative uses (Wiltsee, 1998; McKeever, 2004). One of the challenges facing potential availability of urban woody waste is to sort out the operation that is available (not currently used) and determine alternative uses, including those used by captive markets (not likely to be diverted to bioenergy). One assessment of urban wood waste finds that 36% of total biomass generated is currently solid to noncaptive markets, and 50% of the unused residues are not available due to contamination, quality, or recoverability.

One source of recurring and potentially available carbon feedstock is MSW. In 2005, the USEPA estimated that 245.7 million tons of MSW were generated in the United States, of which 79 million tons were recycled, 33.4 million tons were diverted to energy recovery, and 133.3 million tons were disposed of in landfills. As such, landfilled material represents a potentially significant source of renewable carbon that could be used for fuel/energy production or in support of biofuel production.

ATTRIBUTES OF THE PERFECT ENERGY CROP

The perfect energy crop has

- High biomass
- Improved composition and structure
- Disease and pest resistance
- Optimized architecture
- Salt, pH, and aluminum tolerance
- Rapid and cost-effective propagation
- Stand establishment (cold germination and cold growth)
- Perennial
- Deep roots

SECOND-GENERATION FEEDSTOCKS: LONG-TERM AVAILABILITY

Large-scale biofuels production, in the long run, will require other resources, including dedicated energy crops. Dedicated feedstocks are perennial grasses and trees grown as crops specifically to provide the required raw materials to bioenergy producers. A steady supply of low-cost, uniform, and consistent-quality biomass feed stock will be critical for the economic viability of cellulosic ethanol production. During the late 1980s, the USDOE sponsored research on perennial herbaceous (grassy) biomass crops, particularly switchgrass, considered a model energy crop because of its many perceived advantages: native to North America, high biomass yield per acre, wide regional coverage, and adaptability to marginal land conditions. An extensive research program on switchgrass in the 1990s generated a wealth of information on high-yielding varieties, regional adaptability, and management practices. Preliminary field trials show that the economic viability of switchgrass cultivation depends critically on the initial establishment success. During this phase, seed dormancy and seedling sensitivity to soil and weed conditions require that recommended practices be closely followed by growers. Viable yields require fertilization rates at about half the average for corn.

Switchgrass (*Panicum virgatum*), a hardy, deep-rooted, perennial rhizomatous warm-season grass native to North America, is believed to be that most suitable for cultivation in marginal lands, low-moisture lands, and lands with lower opportunity costs such as pastures, including lands under the Conservation Reserve Program (CRP) where the federal government pays landowners annual rent for keeping land out of production (McLaughlin & Kzos, 2005). Additionally, a large amount of highly erodible land in the Corn Belt is unsuitable for straw or stover removal but potentially viable for dedicated energy crops such as switchgrass. Factors favoring adoption of switchgrass include selection of suitable lands, environmental benefits (carbon balances, improved soil nutrients, and quality), and use of existing hay production techniques to grow the crop. Where switchgrass is grown on CRP lands, payments help offset production costs. Factors discouraging switchgrass adoption include no possibility for crop rotation; farmers' risk aversion for producing a new crop because of lack of information, skills, and know-how; potential conflict with on-farm and off-farm scheduling activities; and a lack of compatibility with long-term land tenure. Overall, production budget and delivery cost assessments suggest that switchgrass is a high-cost crop (undercurrent technology and price conditions) and may not compete with established corps, except in areas with low opportunity costs (e.g., pastureland, marginal lands).

DID YOU KNOW?

According Parrish et al. (2008), regarding establishing switchgrass, no single method can be suggested for all situations. No-till and convention tillage can be used to establish the crop. When seeded as part of a diverse mixture, planting guidelines for warm-season grass mixtures for conservation plantings should be followed. Several key factors can increase the likelihood of success for establishing switchgrass. These include:

- Planting switchgrass after the soil is well warmed during the spring.
- Using seeds that are highly germinal and planting 0.6–1.2 cm deep, or up to 2 cm deep in sandy soils.
- Packing or firming the soil both before and after seeding.
- Providing no fertilization at planting to minimize competition.
- Controlling weeds with chemical and/or cultural control methods.

Substantial variability is apparent in the economics of switchgrass production and assessments of production budgets and delivered costs. Factors at play include methods to store and handle, transport distances, yields, and types of land used (cropland versus grassland). When delivered costs of switchgrass are translated into break-even prices (compared with conventional crops), it becomes apparent that cellulosic ethanol or biopower plants would have to offer relatively high prices for switchgrass to induce farmers to grow it (Rinehart, 2006). However, the economics of switchgrass could improve if growers benefited from CRP payments and other payments tied to environmental services (such as carbon credits). In the long run, the viability of an energy crop like switchgrass hinges on continued reductions in cellulosic ethanol conversion costs and sustained improvements in switchgrass yield and productivity through breeding, biotechnology, and agronomic research.

DID YOU KNOW?

The main advantage of using switchgrass over corn as an ethanol feedstock is that its cost of production is generally about one-half that of grain corn, and more biomass energy per hectare can be captured in the field (Samson et al., 2008).

Although an important biofuels crop, switchgrass does have limitations. Switchgrass is not optimally grown everywhere. For example, in the upper Midwest under wet soils, reed canary grass is more suitable, while semitropical grass species are better adapted to the Gulf Coast region. State and local efforts are testing alternatives to switchgrass such as, as mentioned, a reed canary grass (*Phalaris arundinacea*—a tall, perennial grass that forms extensive single-species stands along the margins of lakes and streams), and *Miscanthus* (often confused with "Elephant Grass," which is a rapid growth, low mineral content, high biomass yield plant), and other species are underway.

Regarding the perspective of long-run sustainability, the ecology of perennial grassy crops favors a multiplicity of crops or even a mix of species within the same area. Both ecological and economic sustainability favor the development of a range of herbaceous species for optimal use of local soil and climatic condition. A mix of several energy crops in the same region would help reduce risk of epidemic pests and disease outbreaks and optimize the supply of biomass to an ethanol or biopower plant since different grasses mature and can be harvested at different times. Moreover, development of future energy corps must be evaluated from the standpoint of their water use efficiency, impact on soil nutrient cycling, effect on crop rotations, and environmental benefit (improved energy use efficiency and reduction in greenhouse gas emissions, nutrient runoff, pesticide runoff, and land-use impacts). In the long run, developing a broad range of grassy corps for energy use is compatible with both sustainability and economic viability criteria.

SHORT-ROTATION WOODY CROPS (SRWC)

Short-rotation woody crops (SRWC) represent another important category of future dedicated energy crops. Among the SRWC, hybrid poplar, willow, American Sycamore, sweetgum, and loblolly have been extensively researched for their very high biomass yield potential. Breeding programs and management practices continue to be developed for these species. SRWC are based on a high-density plantation system and more frequent harvesting (every 3–4 years for willow and 7 years for hybrid poplar).

- *Hybrid poplar*—this species is very site-specific and has a limited growing niche. It requires an abundant and continuous supply of moisture during the growing season. Soils should be moist, but not continually saturated and with good internal drainage. It prefers damp, well-drained, fine sandy-loam soils located near streams, where coarse sand is first deposited as flooding occurs. Normally, cottonwood trees are found at elevations 20 feet above the average level of adjacent streams. Hybrid poplars are among the fastest-growing trees in North America—in just six growing seasons, hybrid poplars reach 60 feet or more in height. They are well suited to produce bioenergy, fiber, and other biased products.
- *Willow Species (Salix)*—in folklore and myth, a willow tree is believed to be quite sinister, capable of uprooting itself and stalking travelers. The reality is willow is used for biofiltration, constructed wetlands, ecological wastewater treatment systems, hedges, land reclamation, landscaping, phytoremediation, stream bank stabilization (bioengineering), slope stabilization, soil erosion control, shelterbelt and windbreak, soil building, soil reclamation, tree bog compost toiler, and wildlife habitat. Willow is grown for biomass or biofuel, in energy forestry systems, because of its high energy in–energy out ratio, large carbon mitigation potential, and fast growth (Aylott, 2008). Willow is also grown and harvested for making charcoal.
- *American Sycamore*—prefers alluvial soils along streams in bottomlands. Sycamore growth and yield are less than cottonwood and willow.
- *Sweetgum (Liquidambar styraciflua)*—is a species tolerant of a variety of soils, but grows best on rich, moist, alluvial clay and loam soils of river bottoms.
- *Loblolly Pine (Pinus taeda)*—is quite adaptable to a variety of sites. It performs well on both poorly drained bottomland flats and modestly arid uplands. Biomass for energy is currently being obtained from precommercial thinning and from logging residue in loblolly pine stands. Utilization will undoubtedly increase, and loblolly pine energy plantations may become a reality.

In many parts of the country, plantations of willow, poplar, pines, and cottonwood have been established and are being commercially harvested. Willow plants are being planted in New York, particularly following the enactment of the Renewable Portfolio Standard (RPS) and other state incentives. Over 30,000 hectares of poplar are grown in Minnesota, and several thousand hectares are also grown as part of a USDOE-funded project to provide biomass for a power utility company in southern

Minnesota. The Pacific Northwest had a large plantation of hybrid poplars, estimated at 60,000 hectares as of 2007. Most of these plantations are currently used for pulp wood, with little volume being used for bioenergy. Since SRWC can be used either for biomass or as feedstock for pulp and other products, pulp demand will influence the cost of using it for bioenergy production.

An important consideration for energy crops (e.g., switchgrass, poplar, willow) is the potential for increasing yield sand developing other desirable characteristics. Most energy crops are unimproved or have been bred only recently for biomass yield, whereas corn and other commercial food crops have undergone substantial improvements in yield, disease resistance, and other agronomic traits. A more complex understanding of biological systems and application of the latest biotechnological advances would accelerate the development of new biomass crops with desirable attributes. These attributes include increased yields and processability, optimal growth in specific microclimates, better pest resistance, efficient nutrient use, and greater tolerance to moisture deficits and other sources of stress. Agronomic and breeding improvements of these new crops could provide a significant boost to future energy crop production.

GASOLINE GALLON EQUIVALENT (GGE)

Before beginning our discussion about alternative renewable fuels used for transportation purposes, it is important to ensure that the reader has a fundamental understanding of the difference between conventional gasoline and diesel fuel energy output as compared to nonconventional renewable products. Typically, this comparison is made utilizing a standard engineering parameter known as the gasoline gallon equivalent (GGE), which is the ratio of the number of British thermal units (BTU) available in one US gallon (1 gal) of gasoline to the number of British thermal units available in 1 gal of the alternative substance in question. In 1994, the National Institute of Standards and Technology (NIST, 1994) defined "gasoline gallon equivalent" (GGE) as 5,660 pounds of natural gas. The GGE parameter allows consumers to compare the energy content of competing fuels against a commonly known fuel—gasoline. Table 5.4 shows the GGE and Btu/unit value comparisons for the listed fuels.

BIOETHANOL PRODUCTION: DRY CORN MILL PROCESS

Ethanol production from dry corn milling follows a seven-step process shown in Figure 6.5 (USDOE, 2010). First, corn feedstock is cleaned and milled (ground into corn meal). Then, milled corn is converted to slurry; the slurry is liquefied (slurried with water—mash), and enzymes yield glucose. Yeast is then added to the mash to convert starch into a simple sugar (dextrose) (saccharification). After liquefaction, the mash is cooked in a saccharification tank to reduce bacterial levels and paled in a fermenter where fermentation takes place (sugar is converted to ethanol by yeast). The resulting "beer" containing 2%–12% ethanol is then distilled into ethanol at 95% alcohol and 5% water, and the remaining solids (stillage) are collected during distillation, dried, and sold as an animal feed called dried distillers' grains (DDG) (see

TABLE 5.4
Gasoline Gallon Equivalents (GGE)

Fuel	GGE	Btu/unit
Gasoline (Base)	1 US Gallon	114,000 Btu/gal
Gasoline (conventional, summer)	0.996 US gallon	114,500 Btu/gal
Gasoline (convention, winter)	1.013 US gallon	112,500 Btu/gal
Gasoline (reformulated gasoline, ethanol	1.019 US gallon	111,836 Btu/gal
Gasoline (reformulated gasoline, ETBE)	1.019 US gallon	111,811 Btu/gal
Gasoline (reformulated gasoline, MTBE)	1.020 US gallon	111,745 Btu/gal
Gasoline (10% MBTE)	1.02 US gallon	112,000 Btu/gal
Gasoline (regular unleaded)	1 US gallon	114,100 Btu/gal
Diesel #2	0.88 US gallons	129,500 Btu/gal
Biodiesel (B100)	0.96 US gallons	118,300 Btu/gal
Bio Diesel (B20)	0.90 US gallons	127,250 Btu/gal
Liquid natural gas (LNG)	1.52 US gallons	75,000 Btu/gal
Compressed natural gas	1.26 cu ft (3.587 m^3)	900 Btu/cu ft
Hydrogen at 101.325 kPa	357.37 cu ft	319 Btu/cu ft
Hydrogen by weight	0.997 kg (2.198 lb)	119.9 MJ/kg (51,500 Btu/lb)
Liquefied petroleum gas (LPG)	1.35 US gallons	84,300 Btu/gal
Methanol fuel (M100)	2.01 US gallons	56,800 Btu/gal
Ethanol fuel (E100)	1.500 US gallons	76,100 Btu/gal
Ethanol (E85)	1.39 US gallons	81,800 Btu/gal
Jet fuel (naphtha)	0.97 US gallons	118,700 Btu/gal

Source: USEPA (2007).

Figure 5.5). The removal of water from ethanol beyond the last 5% is called dehydration or drying.

DID YOU KNOW?

Fermentation is a series of chemical reactions that convert sugars to ethanol. The fermentation reaction is caused by yeast or bacteria, which feed on the sugars. Ethanol and carbon dioxide are produced as the sugar is consumed. The simplified fermentation reaction equation for the 6-carbon sugar, glucose, is:

$$C_6H_{12}O_6 \rightarrow 2\ CH_3CH_2OH + 2\ CO_2$$
glucose   ethanol   carbon dioxide

Corn (1 Bushel)

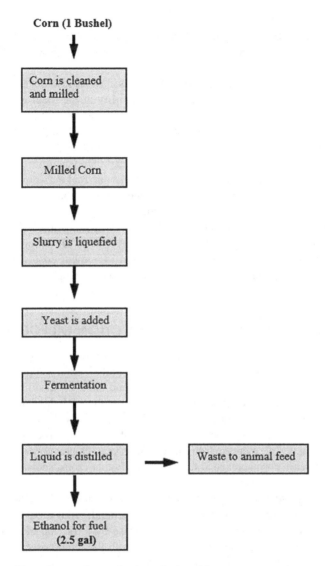

FIGURE 5.5 Flow diagram for production of ethanol from corn.

PROS AND CONS OF BIOETHANOL

Regarding ethanol's use for transportation (propulsion), it can be used as an extender and octane enhancer in conventional gasoline. Its primary advantage is that its feedstock (plants) is renewable. It can also be used as a primary fuel (in E85), thus reducing dependence on petroleum products. Ethanol is an important player in the ongoing effort to reduce environmental pollution from carbon monoxide (CO) gas and global carbon dioxide (CO_2). In some locations, E85 and gasohol are cheaper per gallon than conventional gasoline. As the production and use of ethanol as a fuel increases, farmers will benefit from the increased demand for their products. From the practical and safety and health point of view, ethanol can be used to prevent gas-line freeze in extremely cold weather, and because it is not as flammable as gasoline, it is therefore less likely to cause accidental explosions.

Ethanol use for transportation purposes does have a few negatives. Probably the most obvious shortcoming is its non-availability at the local gas station. Gas stations selling E85 as part of their normal products are not as numerous as conventional gas stations. In some locations, E85 and gasohol are more expensive per gallon than conventional gasoline. E85 can damage vehicles that are not designed to burn it as a fuel. Also, ethanol contains less energy than gasoline. This means that your car won't go as far on a gallon of E85, and your fuel economy will decrease by 20%–30%. Ethanol is produced from plant matter that could otherwise be consumed as human and animal food products. Critics argue that food products used for fuel instead of to feed people and animals will contribute indirectly to mass world hunger. Moreover, it will lead to higher food prices. However, while it is true that higher corn prices increase animal feed and ingredient costs for farmers and food manufacturers, these are passed through to retail at a price rate less than 10% of the corn price change.

SIDEBAR 5.1[3] CORN: ETHANOL PRODUCTION OR FOOD SUPPLY?

In 2007, record US trade driven by economic growth in developing countries and favorable exchange rates, combined with tight global grain supplies, resulted in record or near-record prices for corn, soybeans, and other food and feed grains. For corn, these factors, along with increased demand for ethanol, helped push prices from under $2 per bushel in 2005 to $3.40 per bushel in 2007. By the end of the 2006/07 crop year, over 2 billion bushels of corn (19% of the harvested crop) were used to produce ethanol, a 30% increase from the previous year. Higher corn prices motivated farmers to increase corn acreage at the expense of other crops, such as soybeans and cotton, raising their prices as well.

The pressing and pertinent question thus becomes: What effect do these higher commodity costs have on retail food prices? In general, retail food prices are much less volatile than farm-level prices and tend to rise by a fraction of change in farm prices. The magnitude of response depends on both the retail costs beyond the raw food ingredients and the nature of competition in retail food markets. Ethanol's impact on retail food prices depends on how long the increased demand for corn drives up farm corn prices and the extent to which higher corn prices are passed through to retail.

Retail food prices adjust as the cost of inputs into retail food production changes and the competitive environment in each market evolves. Strong competition among three to five retail store chains in most US markets has had a moderating effect on food price inflation. Overall, retail food prices have been relatively stable over the past 20 years, with prices increasing an average of 3.0% per year from 1987 through 2007, just below the overall rate of inflation. Since then, food price inflation has averaged just 2.5% per year.

Field corn is the predominant corn type grown in the United States, and it is primarily used for animal feed. Currently, less than 10% of the US field corn crop is used for direct domestic human consumption in corn-based foods such as cornmeal, cornstarch, and cornflakes, while the remainder is used for animal feed, exports, ethanol production, seed, and industrial uses. Sweet corn, both white and yellow, is usually consumed as immature whole-kernel corn by humans and as an ingredient in other corn-based foods, but makes up only about 1% of the total US corn production.

Because US ethanol production uses field corn, the most direct impact of increased ethanol production should be on field corn prices and on the price of food products based on field corn. However, even for those products heavily based on field corn, the effect of rising corn prices is dampened by other market factors. For example, an 18-ounce box of cornflakes contains about 12.9 ounces of milled field corn. When field corn is priced at $2.28 per bushel (the 20-year average), the actual value of corn represented in the box of cornflakes is about 3.3 cents (1 bushel = 56 pounds). (The remainder is packaging, processing, advertising, transportation, and other costs.) At $3.40 per bushel, the average price in 2007, the value was about 4.9 cents. The 49% increase in corn prices would be expected to raise the price of a box of cornflakes by about 1.6 cents, or 0.5%, assuming no other cost increases.

In 1985, Coca-Cola made the shift from sugar to corn syrup in most of its US-produced soda, and many other beverage makers followed suit. Currently, about 4.1% of US-produced corn is made into high-fructose corn syrup. A 2-liter bottle of soda contains about 15 ounces of corn in the form of high-fructose corn syrup. At $3.40 per bushel, the actual value of corn represented is 5.7 cents, compared with 3.8 cents when corn is priced at $2.28 per bushel. Assuming no other cost increases, the higher corn price in 2007 would be expected to raise soda prices by 1.9 cents per 2-liter bottle, or 1%. These are notable changes in terms of price measures and inflation, but relatively minor changes in the average household budget.

In addition to the impact on cornflakes and soda beverage prices, livestock prices are also impacted. This stands to reason when you consider that livestock rations traditionally contain a large amount of corn; thus, a bigger impact would be expected in meat and poultry prices due to higher feed costs than in other food products. Currently, 55% of corn produced in the United States is used as animal feed for livestock and poultry. However, estimating the actual corn used as feed to produce retail meat is a complicated calculation. Livestock producers have many options when deciding how much corn to include in a feed ration. For example, at one extreme, grass-fed cattle consume no corn, while other cattle may have a diet consisting primarily of corn. For hog and poultry producers, ration variations may be less extreme, but can still vary quite a bit. To estimate the impact of higher corn prices on retail meat prices, it is necessary to make a series of assumptions about feeding practices and grain conversion rates from animal to final retail meat products. To avoid downplaying potential impacts, this analysis uses upper-bound conversion estimates of 7 pounds of corn to produce 1 pound of beef, 6.5 pounds of corn to produce 1 pound of pork, and 2.6 pounds of corn to produce 1 pound of chicken.

Using these ratios and data from the Bureau of Labor Statistics, a simple pass-through model provides estimates of the expected increase in meat prices given the higher corn prices. The logic of this model is illustrated by an example using chicken prices. Over the past 30 years, the average price of a bushel of corn in the United States has been $2.28, implying that a pound of chicken at the retail level uses 8 cents worth of corn, or about 4% of the $2.05 average retail price for chicken breasts. Using the average price of corn for 2007 ($3.40 per bushel) and assuming producers do not change their animal-feeding practices, retail chicken prices would rise 5.2 cents, or 2.5%. Using the same corn data, retail beef prices would go up 14 cents per pound, or 8.7%, while pork prices would rise 13 cents per pound, or 4.1%.

These estimates for meat, poultry, and corn-related foods, however, assume that the magnitude of the corn price change does not affect the rate at which cost increases are passed through to retail prices. It could be the case that corn price fluctuations have little impact on retail food prices until corn prices rise high enough for a long enough time to elicit a large price adjustment by food producers and notably higher retail food prices.

On the other hand, these estimates may be overstating the effect of corn price increases on retail food prices since they do not account for potential substitution by producers from more expensive to less costly inputs. Such substitution would dampen the effect of higher corn prices on retail meat prices. Even assuming the upper-bound effects outlined above, the impact of rising corn commodity prices on overall food prices is limited. Given that less than a third of retail food contains corn as a major ingredient, these rising prices for corn-related products would raise overall US retail food prices less than 1 percentage point per year above the normal rate of inflation.

Continuing elevated prices for corn will depend on the extent to which corn remains the most efficient feedstock for ethanol production and ethanol remains a viable source of alternative energy. Both conditions may change over time as other crops and biomass are used to produce ethanol and other alternative energy sources develop.

DID YOU KNOW?

While higher commodity costs may have a relatively modest impact on US retail food prices, there may be greater effect on retail food prices in lower-income developing countries. As a relatively low-priced food, grains have historically accounted for a large share of the diet in less developed countries. Even with incomes rising, consumers in such countries consume a less processed diet than is typical in the United States and other industrialized countries, so food prices are more closely tied to swings in both domestic and global commodity prices.

Even if these conditions do not change in the near term, market adjustments may dampen long-run impacts. In 1996, when field corn prices reached an all-time high of $3.55 per bushel due to drought-related tighter supplies in the United States and strong demand for corn from China and other parts of Asia, the effect on food prices was short lived. At that time, retail prices rose for some foods, including pork and poultry, but these effects did not extend beyond the middle of 1997. For the most part, food markets adjusted to the higher corn prices and corn producers increased supply, bringing down prices.

Food producers, manufacturers, and retailers may also adjust to the changing market conditions by adopting more efficient production methods and improved technologies to counter higher costs. For example, soft drink manufacturers may consider substituting sugar for corn syrup as a sweetener if corn prices remain high, while livestock and poultry producers may develop alternative feed rations that minimize corn needed for animal feed. Adjustments by producers, manufacturers, and retailers, along with continued strong retail competition, suggest that US retail food prices will remain relatively stable.

BIODIESEL

The diesel engine is the workhorse of heavy transportation and industrial processes; it is widely used to power trains, tractors, ships, pumps, and generators. Powering the diesel engine is conventional diesel fuel or biodiesel fuel. Biodiesel is a rather viscous liquid fuel made up of fatty acid alkyl esters, fatty acid methyl esters (FAME), or long-chain mono alkyl esters. It is produced from renewable sources such as new and used vegetable oils and animal fats and is a cleaner-burning replacement for petroleum-based diesel fuel. It is nontoxic and biodegradable. This fuel is designed to be used in compression-ignition (diesel) engines similar or identical to those that burn petroleum diesel. Biodiesel has physical properties (see Table 5.5) like those of petroleum diesel.

In the United States, most biodiesel is made from soybean oil or recycled cooking oils. Animal fats, other vegetable oils, and other recycled oils can also be used to produce biodiesel, depending on the costs and availability. In the future, blends of all kinds of fats and oils may be used to produce biodiesel.

Before providing a basic description of the vegetable and/or recycled grease to biodiesel process, it is important to define and review a few of the key technological terms associated with the conversion process.

TABLE 5.5
Biodiesel Physical Characteristics

Specific gravity	0.87 to 0.89
Kinematic viscosity @ 40°C	3.7 to 5.8
Cetane number	46 to 70
Higher heating value (btu/lb)	16,928 to 17,996
Sulfur, wt%	0.0 to 0.0024
Cloud point °C	−11 to 16
Pour point °C	−15 to 13
Iodine number	60 to 135
Lower heating value (lbs/lb)	15,700 to 16,735

- *Acid Esterification*—oil feedstocks containing more than 4% free fatty acids go through an acid esterification process to increase the yield of biodiesel. These feedstocks are filtered and preprocessed to remove water and contaminants, and then fed to the acid esterification process. The catalyst, sulfuric acid, is dissolved in methanol and then mixed with the pretreated oil. The mixture is heated and stirred, and the free fatty acids are converted to biodiesel. Once the reaction is complete, it is dewatered and then fed to the transesterification process.

- *Transesterification*—oil feedstocks containing less than 4% free fatty acids are filtered and preprocessed to remove water and contaminants and then fed directly to the transesterification process along with any products of the acid esterification process. The catalyst, potassium hydroxide, is dissolved in methanol and then mixed with and the pretreated oil. If an acid esterification process is used, then extra base catalyst must be added to neutralize the acid added in that step. Once the reaction is complete, the major co-products, biodiesel and glycerin, are separated into two layers.

- *Methanol recovery*—the methanol is typically removed after the biodiesel and glycerin have been separated, to prevent the reaction from reversing itself. The methanol is cleaned and recycled back to the beginning of the process.

- *Biodiesel refining*—once separated from the glycerin, the biodiesel goes through a cleanup or purification process to remove excess alcohol, residual catalyst, and soaps. This consists of one or more washings with clear water. It is then dried and sent to storage. Sometimes the biodiesel goes through an additional distillation step to produce a colorless, odorless, zero-sulfur biodiesel.

- *Glycerin refining*—the glycerin by-product contains unreacted catalyst and soaps that are neutralized with an acid. Water and alcohol are removed to produce 50%–80% crude glycerin. The remaining contaminants include unreacted fats and oils. In large biodiesel plants, the glycerin can be further purified, to 99% or higher purity, for sale to the pharmaceutical and cosmetic industries.

The most popular biodiesel production process is transesterification (production of the ester) of vegetable oils or animal fats, using alcohol in the presence of a chemical catalyst. About 3.4 kg of oil/fat is required for each gallon of biodiesel produced (Baize, 2006). The transesterification of degummed soybean oil produces ester and glycerin. The reaction requires heat and a strong base catalyst such as sodium hydroxide or potassium hydroxide. The simplified transesterification reaction is shown below.

base

Triglycerides + Free Fatty acids (<4%) + Alcohol $\rightarrow$ Alkyl esters + glycerin

Some feedstocks must be pretreated before they can go through the transesterification process. Feedstocks with less than 4% free fatty acids, which include vegetable oils and some food-grade animal fats, do not require pretreatment. Feedstocks with more than 4% free fatty acids, which include inedible animal fats and recycled greases, must be pretreated in an acid esterification process. In this step, the feedstock is reacted with an alcohol (like methanol) in the presence of a strong acid catalyst (sulfuric acid), converting the free fatty acids into biodiesel. The remaining triglycerides are converted to biodiesel in the transesterification reaction.

acid

$$\text{Triglycerides} + \text{Free Fatty Acids} (>4\%) + \text{Alcohol} \rightarrow \text{Alkyl esters} + \text{triglycerides}$$

Figure 5.6 illustrates the basic technology for processing vegetable oils (such as soybeans) and recycled greases (used cooking oil and animal fat). When the feedstock is vegetable oil, the extracted oil is processed to remove all traces of water, dirt, and other contaminants. Free fatty acids are also removed. A combination of methyl alcohol and a catalyst, usually sodium hydroxide or potassium hydroxide, breaks the oil molecules apart in the esterification process. The resulting esters are then refined into usable biodiesel.

When the feedstock is used-up cooking oil and animal fats refined to produce biodiesel, the process is like the way biodiesel is derived from vegetable oil, except there is an additional step involved (Figure 5.6). Methyl alcohol and sulfur are used in a process called dilute acid esterification to obtain a substance resembling fresh vegetable oil, which is then processed in the same way as vegetable oil to obtain the final product.

DID YOU KNOW?

Biodiesel is routinely blended with petroleum diesel. The percentage of biodiesel in the blend is written following an uppercase letter "B" to denote the proportion. For example, a mixture of 20% biodiesel and 80% petroleum diesel is called B20; a mixture of equal parts biodiesel and petroleum diesel is B50; and neat (or pure) biodiesel is B100. Most conventional diesel engines can burn blends from pure petroleum diesel up to B20 without modification. With minor modification, most diesel engines built since 1994 can use blends from B20 to B100. The diesel vehicle owner's manual and vehicle warranty should be checked before using any alternative fuel.

ALGAE TO BIODIESEL

The focus of our discussion of biofuels to this point has been on terrestrial sources of biomass fuels. This section is concerned with photosynthetic organisms that grow in aquatic environments, namely, macroalgae, microalgae, and emergent. Macroalgae, more commonly known as "seaweed," are fast-growing marine and freshwater plants that can grow to considerable size (up to 60 m in length). Emergent are plants that grow partially submerged in bogs and marshes.

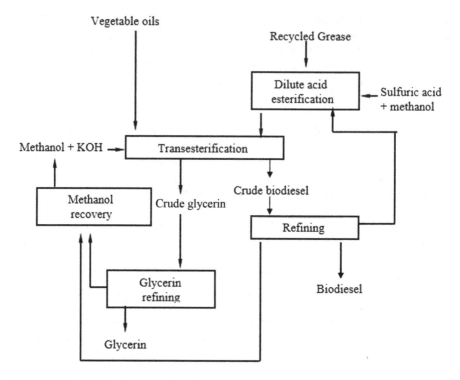

FIGURE 5.6   Biodiesel: process description.

Regarding the mass production of oil for alternative renewable energy purposes, the focus of this discussion is mainly on microalgae, organisms capable of photosynthesis that are less than 0.4 mm in diameter, including diatoms and cyanobacteria, as opposed macroalgae (as mentioned, seaweed). Microalgae are the preferred choice because they are less complex in structure, have a fast growth rate, and typically (for some species) contain a high concentration of oil. However, it is important to point out that recent research efforts are also focusing on using seaweed for algae fuel (biofuels), probably due to their high availability (Time Online, 2005).

DID YOU KNOW?

Algae fuel, also called algal fuel, algae oleum, or second-generation biofuel, is a biofuel that is derived from algae, that is, the production of natural oils for biodiesel.

The following algal species listed are currently being studied for their suitability as a mass-oil producing crop, across various locations worldwide (USDOE, 2010; Taipei Times, 2008).

Ulva
Botryococcus braunii
Chlorella
Dunaliella tertiolecta
Gracilaria
Pleurochrysis carterae
Sargassum (has 10 times the output volume of Gracilaria)

The algae-to-biodiesel alternative renewable fuel source program got its start during the Carter administration in response to the 1970s Arab fuel embargo. At the same time, the Carter administration consolidated all federal energy activities under the auspice of the newly established USDOE. Among its various programs established to develop all forms of alternative energy (related to solar energy), the USDOE initiated research on the use of plant life as a source of transportation fuels (today called the Biofuels Program).

Before discussing the algae-to-biofuels process in detail, it is important for the reader to be well grounded in basic algal concepts. In many ways, the study of microalgae is a relatively limited field of study. Algae are not nearly as well understood as other organisms that have found a role in today's biotechnology industry. The study of microalgae represents an area of higher risk and high gains; it also presents an opportunity for the curious and ambitious to break new ground in science by conducting research in this growing area of interest.

DID YOU KNOW?

In the 1980s, the USDOE gradually shifted its focus to technologies that could have large-scale impacts on national consumption of fossil energy. Many of the USDOE's publications from this period reflect a philosophy of energy research that might, somewhat pejoratively, be called "the quads mentality." A quad is a shorthand name for the unit of energy often used by the USDOE to describe the amounts of energy that a given technology might be able to displace. Quad is short for "quadrillion Btus"—a unit of energy representing 10^{15} (1,000,000,000,000,000) Btus of energy. This perspective led the USDOE to focus on the concept of immense algae farms.

Algae

The protists that perform photosynthesis are called **algae**. Algae can be both a nuisance and an ally; we might say they possess "Jekyll and Hyde-like" characteristics (good vs. bad characteristics). Many ponds, lakes, rivers, streams, and bays (e.g., Chesapeake Bay) in the United States (and elsewhere) are undergoing **eutrophication** (basically the killing off, although in many cases very slowly, of calm water environments, especially ponds, marshes, and lakes), or the enrichment of an environment with inorganic substances (phosphorous and nitrogen). When eutrophication occurs, when filamentous algae like *Caldophora* break loose in a pond, lake, stream, or river and washes ashore, algae make it stink and makes their noxious presence known. More importantly than the offensive odor dying algae give off is their deaths start the biodegradation process whereby, instead of adding oxygen to their watery environment, they begin to use it up. As mentioned, algae have a good side too. For instance, algae are allies in many wastewater treatment operations. In addition, they can be valuable in long-term oxidation ponds where they aid in the purification process by producing oxygen.

Before discussing the specifics and different types of algae, it is important to be familiar with algal terminology.

Algal Terminology

- **Algae**—large and diverse assemblages of eukaryotic organisms that lack roots, stems, and leaves but have chlorophyll and other pigments for carrying out oxygen-producing photosynthesis.
- **Algology** or **Phycology**—the study of algae.
- **Antheridium**—special male reproductive structures where sperm are produced.
- **Aplanospore**—nonmotile spores produced by sporangia.
- **Benthic**—algae attached and living on the bottom of a body of water.
- **Binary fission**—nuclear division followed by division of the cytoplasm.
- **Chloroplasts**—packets that contain *chlorophyll a* and other pigments.
- **Chrysolaminarin**—the carbohydrates reserved in organisms of division **Chrysophyta**.
- **Diatoms**—photosynthetic, circular, or oblong chrysophyte cells.
- **Dinoflagellates**—unicellular, photosynthetic protistan algae.
- **Dry mass factor**—the percentage of dry biomass in relation to the fresh biomass; for example, if the dry mass factor is 5%, one would need 20 kg of wet algae (algae in the media) to get 1 kg of dry algae cells.
- **Epitheca**—the larger part of the frustule (diatoms).
- **Euglenoids**—contain chlorophylls **a** and **b** in their chloroplasts; representative genus is *Euglena*.
- **Fragmentation**—a type of asexual algal reproduction in which the thallus breaks up and each fragmented part grows to form a new thallus.
- **Frustule**—the distinctive two-piece wall of silica in diatoms.
- **Hypotheca**—the small part of the frustule (diatoms).
- **Lipid content**—the percentage of oil in relation to the dry biomass needed to get it, that is, if the algae lipid content is 40%, one would need 2.5 kg of dry algae to get 1 kg of oil.
- **Neustonic**—algae that live at the water–atmosphere interface.
- **Oogonia**—vegetative cells that function as female sexual structures in the algal reproductive system.
- **Pellicle**—a *Euglena* structure that allows for turning and flexing of the cell.
- **Phytoplankton**—made up of algae and small plants.
- **Plankton**—free-floating, mostly microscopic aquatic organisms.

- **Planktonic**—algae suspended in water as opposed to attached and living on the bottom (benthic).
- **Protothecosis**—a disease in humans and animals caused by the green algae, *Prototheca moriformis*.
- **Thallus**—the vegetative body of algae.

Algae are autotrophic, contain the green pigment chlorophyll, and are a form of aquatic plant. Algae differ from bacteria and fungi in their ability to carry out photosynthesis—the biochemical process requiring sunlight, carbon dioxide, and raw mineral nutrients. Photosynthesis takes place in the chloroplasts. The chloroplasts are usually distinct and visible. They vary in size, shape, distribution, and number. In some algal types, the chloroplast may occupy most of the cell space. They usually grow near the surface of water because light cannot penetrate very far through water. Although in mass (multicellular forms like marine kelp) the unaided eye easily sees them, many of them are microscopic. Algal cells may be nonmotile, motile by one or more flagella, or exhibit gliding **motility** as in diatoms. They occur most commonly in water (fresh and polluted water, as well as in salt water), in which they may be suspended (Planktonic) phytoplankton or attached and living on the bottom (benthic). A few algae live at the water–atmosphere interface and are termed neustonic. Within the fresh and saltwater environments, they are important primary producers (at the start of the food chain for other organisms). During their growth phase, they are important oxygen-generating organisms and constitute a significant portion of the plankton in water.

According to the five-kingdom system of Whittaker, algae belong to seven divisions distributed between two different kingdoms. Although seven divisions of algae occur, only five divisions are discussed in this text:

- **Chlorophyta**—Green algae
- **Euglenophyta**—Euglenids
- **Chrysophyta**—golden-brown algae, diatoms
- **Phaeophyta**—Brown algae
- **Pyrrophyta**—Dinoflagellates

The primary classification of algae is based on cellular properties. Several characteristics are used to classify algae, including: (1) cellular organization and cell wall structure; (2) the nature of chlorophyll(s) present; (3) the type of motility, if any; (4) the carbon polymers that are produced and stored; and (5) the reproductive structures and methods. Table 5.6 summarizes the properties of the five divisions discussed in this text.

Algae show considerable diversity in the chemistry and structure of their cells. Some algal cell walls are thin, rigid structures usually composed of cellulose modified by the addition of other polysaccharides. In other algae, the cell wall is strengthened by the deposition of calcium carbonate. Other forms have chitin present in the cell wall. Complicating the classification of algal organisms are the Euglenids, which lack cell walls. In diatoms the cell wall is composed of silica. The frustules (shells) of

TABLE 5.6
Comparative Summary of Algal Characteristics

Algal Group	Common Name	Structure	Pigments	Carbon Reserve	Motility	Reproduction
Chlorophyta	Green algae	Unicellular to Multicellular	Chlorophylls a and b, carotenes, xanthophylls	Starch, oils	Most are nonmotile	Asexual and sexual
Euglenophyta	Euglenoids	Unicellular	Chlorophylls a and b, carotenes, xanthophylls	Fats	Motile	Asexual
Chrysophyta	Golden brown algae, diatoms	Multicellular	Chlorophylls a and b, special carotenoids, xanthophylls	Oils	Gliding by diatoms; others by flagella	Asexual and sexual
Phaeophyta	Brown algae	Unicellular	Chlorophylls a and b, carotenoids xanthophylls	Fats	Motile	Asexual and sexual
Pyrrophyta	Dinoflagellated	Unicellular	Chlorophylls A and b, Carotenes xanthophylls	Starch,	Motile	Asexual; sexual rare

diatoms have extreme resistance to decay and remain intact for long periods of time, as the fossil records indicate.

The principal feature used to distinguish algae from other microorganisms (for example, fungi) is the presence of chlorophyll and other photosynthetic pigments in algae. All algae contain chlorophyll *a*. Some, however, contain other types of chlorophyll. The presence of these additional chlorophylls is characteristic of a particular algal group. In addition to chlorophyll, other pigments encountered in algae include fucoxanthin (brown), xanthophylls (yellow), carotenes (orange), phycocyanin (blue), and phycoerythrin (red).

Many algae have flagella (a threadlike appendage). As mentioned, the flagella are locomotor organelles that may be the single polar or the multiple polar type. The *Euglena* is a simple flagellate form with a single polar flagellum. Chlorophyta have either two or four polar flagella. Dinoflagellates have two flagella of different lengths. In some cases, algae are nonmotile until they form motile gametes (a haploid cell or nucleus) during sexual reproduction. Diatoms do not have flagella but have gliding motility.

Algae can be either autotrophic or heterotrophic. Most are photoautotrophic; they require only carbon dioxide and light as their principal source of energy and carbon. In the presence of light, algae carry out oxygen-evolving photosynthesis; in the absence of light, algae use oxygen. Chlorophyll and other pigments are used to absorb light energy for photosynthetic cell maintenance and reproduction. One of the key characteristics used in the classification of algal groups is the nature of the reserve polymer synthesized because of utilizing carbon dioxide present in water.

Algae may reproduce either asexually or sexually. Three types of asexual reproduction occur: binary fission, spores, and fragmentation. In some unicellular algae, binary fission occurs where the division of the cytoplasm forms new individuals like the parent cell following nuclear division. Some algae reproduce through spores. These spores are unicellular and germinate without fusing with other cells. In fragmentation, the thallus breaks up and each fragment grows to form a new thallus.

Sexual reproduction can involve the tunion of cells where eggs are formed within vegetative cells called oogonia (which function as female structures) and sperm are produced in a male reproductive organ called antheridia. Algal reproduction can also occur through a reduction of chromosome number and/or the union of nuclei.

Algae are photosynthetic organisms that are far from monolithic. Biologists have categorized microalgae in a variety of classes, mainly distinguished by the pigmentation, life cycle, and basic cellular structure.

Characteristics of Algal Divisions

- **Chlorophyta** (Green Algae)—most algae found in ponds belong to this group; they also can be found in salt water and soil. Several thousand species of green algae are known today. Many are unicellular; others are multicellular filaments or aggregated colonies. The green algae have chlorophylls a and b, along with specific carotenoids, and they store carbohydrates in starch. Few green algae are found at depths greater than 7–10 meters, largely because sunlight does not penetrate to that depth. Some species have a holdfast structure that anchors them to the bottom of the pond and to other submerged inanimate objects. Green algae reproduce by both sexual and asexual means. Multicellular green algae have some division of labor, producing various reproductive cells and structures.

- **Euglenophyta** (Euglenoids)—are a small group of unicellular microorganisms that have a combination of animal and plant properties. Euglenoids lack a cell wall, possess a gullet, can ingest food, can assimilate organic substances, and, in some species, are absent of chloroplasts. They occur in fresh, brackish, and salt waters, and on moist soils. A typical *Euglena* cell is elongated and bound by a plasma membrane; the absence of a cell wall makes them very flexible in movement. Inside the plasma membrane is a structure called the pellicle that gives the organisms a definite form and allows the cell to turn and flex. Euglenoids that are photosynthetic contain chlorophylls a and b, and they always have a red eyespot (**stigma**) that is sensitive to light (photoreceptive). Some euglenoids move about by means of flagellum; others move about by means of contracting and expanding motions. The characteristic food supply for euglenoids is a lipopolysaccharide. Reproduction in euglenoids is by simple cell division.
- **Interesting Point**: Some autotrophic species of *Euglena* become heterotrophic when light levels are low.
- **Chrysophyta** (Golden Brown Algae)—the Chrysophycophyta group is quite large—several thousand diversified members. They differ from green algae and euglenoids in that: (1) chlorophylls *a* and *c* are present, (2) fucoxanthin, a brownish pigment, is present; and (3) they store food in the form of oils and leucosin, a polysaccharide. The combination of yellow pigments, fucoxanthin, and chlorophylls causes most of these algae to appear golden-brown. The Chrysophycophyta is also diversified in cell wall chemistry and flagellation. The division is divided into three major classes: golden-brown algae, yellow-brown algae, and diatom.

  Some Chrysophyta lack cell walls; others have intricately patterned coverings external to the plasma membrane, such as walls, plates, and scales. The diatoms are the only group that has hard cell walls of pectin, cellulose, or silicon, constructed in two halves (the epitheca and the hypotheca) called a frustule. Two anteriorly attached flagella are common among Chrysophyta; others have no flagella.

  Most Chrysophyta are unicellular or colonial. Asexual cell division is the usual method of reproduction in diatoms; other forms of Chrysophyta can reproduce sexually.

  Diatoms have direct significance for humans. Because they make up most of the phytoplankton of the cooler ocean parts, they are the ultimate source of food for fish.

  Water and wastewater operators understand the importance of their ability to function as indicators of industrial water pollution. As water quality indicators, their specific tolerances to environmental parameters such as pH, nutrients, nitrogen, concentration of salts, and temperature have been compiled.

- **Interesting Point**: Diatoms secrete a silicon dioxide shell (frustule) that forms the fossil deposits known as diatomaceous earth, which is used in filters and as abrasives in polishing compounds.
- **Phaeophyta** (Brown Algae)—except for a few freshwater species, all algal species of this division exist in marine environments as seaweed. They are a highly specialized group, consisting of multicellular organisms that are sessile (attached and not free-moving). These algae contain essentially the same pigments seen in the golden-brown algae, but they appear brown because of the predominance of and the masking effect of a greater amount of fucoxanthin. Brown algal cells store food such as the carbohydrate laminarin and some lipids. Brown algae reproduce asexually. Brown algae are used in foods, animal feeds, and fertilizers and as source for alginate, a chemical emulsifier added to ice cream, salad dressing, and candy.
- **Pyrrophyta** (Dinoflagellates)—the principal members of this division are the dinoflagellates. The dinoflagellates comprise a diverse group of biflagellated and non-flagellated unicellular, eukaryotic organisms. The dinoflagellates occupy a variety of aquatic environments, with the majority living in marine habitats. Most of these organisms have a heavy cell wall composed of cellulose-containing plates. They store food such as starch, fats, and oils. These algae have chlorophylls *a* and *c* and several xanthophylls. The most common form of reproduction in dinoflagellates is by cell division, but sexual reproduction has also been observed.

DID YOU KNOW?

Cell division in dinoflagellates differs from most protistans, with chromosomes attaching to the nuclear envelope and being pulled apart as the nuclear envelope stretches. During cell division in most other eukaryotes, the nuclear envelope dissolves.

Algal biomass contains three main components:

- Carbohydrates
- Protein
- Natural Oils

Biodiesel production applies exclusively to the natural oil fraction, the main product of interest to us in this section.

The bulk of natural oil made by oilseed crops is in the form of triacylglycerols (TAGs). TAGs consist of three long chains of fatty acids attached to a glycerol backbone. The algae species of concern can produce up to 60% of their body weight in the form of TAGs. (*Note:* Recall that the species of concern, that is, the oil producers, are *Ulva, Botryococcus braunii, Chlorella, Dunaliella tertiolecta, Gracilaria,*

Pleurochrysis carterae, and *Sargassum*). Thus, algae represent an alternative source of biodiesel, one that does not compete with the exiting oilseed market.

Algae-to-Biodiesel Production

Christi (2007) points out that algae can produce up to 300 times more oil per acre than conventional crops, such as rapeseed, palms, soybean, or jatropha. Moreover, algae has a harvesting cycle of 1–10 days, permitting several harvests in a very short time frame, a differing strategy to yearly crops. Algae can also be grown on land that is not suitable for other established crops, for instance, arid land, land with excessively saline soil, and drought-stricken land. This minimizes the issue of taking away pieces of land from the cultivation of food crops (Schenk et al., 2008). Algae can grow 20 to 30 times faster than food crops (McDill, 2010).

Algae can be produced and harvested for biofuel using various technologies. These include photobioreactors (plastic tubes full of nutrients exposed to sunlight), closed-loop (not exposed to open air) systems, and open ponds. For illustration (even though there are many objectors and dissenters to open pond systems), we choose to discuss the open pond types configured in algae farms, because the open pond raceway system is a relatively low-cost system and is an easily understood process.

Open-Pond Algae Farm—Algae farms consist of open, shallow ponds in which some source of waste carbon dioxide (CO_2) can be bubbled into the ponds and captured by the algae (see Figure 5.7).

As shown in Figure 5.7, the ponds in an algae farm are "raceway" designs, in which the algae, water, and nutrients circulate around a racetrack. Paddlewheels provide the flow. The algae are thus kept suspended in water. Algae are circulated back up to the surface on a regular frequency. The ponds are kept shallow because of the

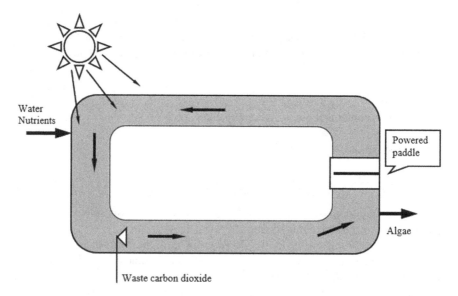

FIGURE 5.7   Raceway design algae pool.

need to keep the algae exposed to sunlight and the limited depth to which sunlight can penetrate the pond water. The ponds operate continuously; that is, water and nutrients are constantly fed to the pond, while algae-containing water is removed at the other end. Some kind of harvesting system is required to recover the algae, which contains substantial amounts of natural oil.

Figure 5.8 illustrates the concept of an "algae farm." The size of these ponds is measured in terms of surface area (as opposed to volume), because surface area is so critical to capturing sunlight. Their productivity is measured in terms of biomass produced per day per unit of available surface area. Even at levels of productivity that would stretch the limits of an aggressive research and development program, such a system will require acres of land. At such large sizes, it is more appropriate to think of these operations on the scale of a farm.

Waste CO_2 is readily available from several sources. Every operation that involves combustion of fuel for energy is a potential source. Generally, coal and other fossil fuel-fired power plants are targeted as the main sources of CO_2. Typical coal-fired

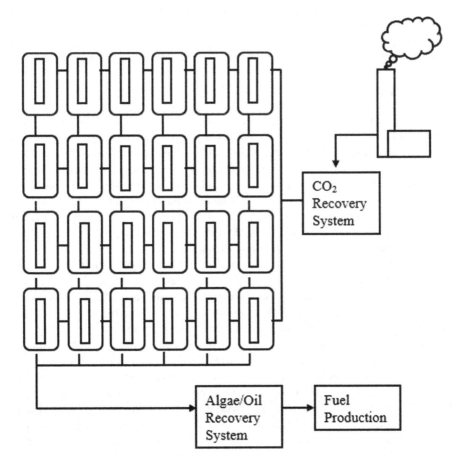

FIGURE 5.8 Algae farm.

power plants emit flue gas from their stacks containing up to 13% CO_2. This high concentration of CO_2 enhances transfer and uptake of CO_2 in the ponds. The concept of coupling a coal-fired power plant with an algae farm provides a win-win approach to recycle the CO_2 from coal combustion into a usable liquid fuel.

JATROPHA TO BIODIESEL

The uninformed or misinformed might flinch when they discover that a by-product of the plant *Jatropha curcas*—yes, the same plant known in some places as "black vomit nut" and in others as "bellyache bush" or "tuba-tub"—is being used as a product with any credible value to anyone. The Jatropha family of shrubs is nonedible plants that grow mostly in countries like the Philippines. It is resistant to drought and can easily be planted or propagated through seeds or cuttings. It starts producing seeds within 14 months but reaches its maximum productivity level after 4–5 years. The Jatropha plant can produce an oil content of 30%–58%, depending on the quality of the soil where it is planted. The seeds yield an annual equivalent of 0.75 to 2 tons, or 1,892 liters of biodiesel per hectare. The plant remains useful for around 30–40 years and can be planted in harsh climates where it won't compete for resources needed to grow food. Along with having the advantage of being a renewable fuel source, the Jatropha plant also reduces greenhouse gas emissions and our dependence on oil imports.

PROS AND CONS OF BIODIESEL

The greatest advantage of biodiesel over conventional petroleum diesel is that biodiesel comes from renewable resources. The supply can be grown, repeatedly. Biodiesel combustion also produces fewer emissions (except for nitrous oxides) than combustion of an equal amount of petroleum diesel. The widespread use of biodiesel can also reduce the dependency on imported oil. From a safety standpoint, biodiesel is safer than petroleum diesel because it is less combustible. From an environmental standpoint, when accidentally spilled, biodiesel is not persistent within environmental media (air, water, soil) because it is biodegradable. Biodiesel can also be produced from waste products such as cooking oils and grease.

Probably the most pressing disadvantages or shortcomings of biodiesel, at least at the present time, are its lack of availability or accessibility and high-cost relative to petroleum diesel. This trend in non-availability and non-accessibility is bound to change as cheaper, more accessible petroleum diesel becomes more expensive and difficult to find. Additional disadvantages to consider are that biodiesel requires special handling, storage, and transportation management as compared to petroleum diesel. Regarding environmental considerations, when combusted, biodiesel produces more nitrous oxide emissions than an equal amount of petroleum diesel. Another potential problem with the production of biodiesel is its dependence on soybeans as its primary feedstock; there is some concern that the widespread use of biodiesel as fuel will contribute to higher food prices and indirectly to world hunger. There is a slight reduction in performance and mileage per gallon with biodiesel as compared with petroleum diesel. Biodiesel can also act as a solvent in some diesel engines, causing loosened deposits that may clog filters.

BIOGAS (METHANE CH$_4$)

Primarily known as a fuel for interior heating systems, methane, or biogas, can also be used as a replacement for natural gas—a fossil fuel for electricity generation and for cooking and heating—and as an alternative fuel to gasoline. Methane is a natural gas produced by the breakdown of organic material in the absence of oxygen in termite mounds, wetlands, and by some animals. Humans are also responsible for the release of methane through biomass burning, rice production, cattle, and release from gas exploration. Methane can also be obtained directly from the Earth; however, other methods of production have been developed, most notably the fermentation or composting of plant and animal waste.

The reasons for considering biogas (methane) as a possible biofuel include:

- It is viable because of its potential use as an alternative fuel source.
- It is a viable alternative fuel to use to improve air quality.
- It can be produced locally reducing the need to use imported natural gas.

Methane is produced under anaerobic (no oxygen) conditions where organic material is biodegraded or broken down by a group of microorganisms. The three main sources of feedstock material for anaerobic digestion are shown in Figure 5.9; these are described in the following sections.

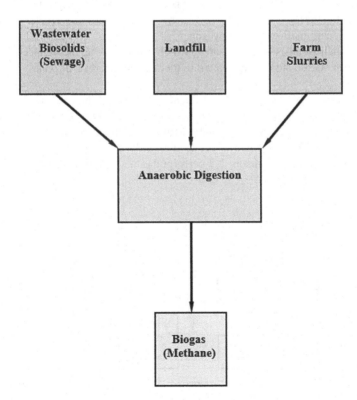

FIGURE 5.9 Production of biogas (methane CH$_4$).

ANAEROBIC DIGESTION

Anaerobic digestion is the traditional method of managing waste, of sludge stabilization, and/or to release energy. It involves using bacteria that thrive in the absence of oxygen and is slower than aerobic digestion but has the advantage that only a small percentage of the waste is converted into new bacterial cells. Instead, most of the organics are converted into carbon dioxide and methane gas.

Cautionary Note: In an anaerobic digester, the entrance of air should be prevented because of the potential for air mixed with the gas produced in the digester, which could create an explosive mixture.

STAGES OF ANAEROBIC DIGESTION

USEPA (1979, 2006) and Spellman (2009) point out that there are four key biological and chemical stages of anaerobic digestion (see Figure 5.10):

1. Hydrolysis—occurs during which the proteins, cellulose, lipids, and other complex organics are broken down into smaller molecules and become soluble by utilizing water to split the chemical bonds of the substances
2. Acidogenesis—occurs during which the products of hydrolysis are converted into organic acids (where monomers are converted to fatty acids)
3. Acetogenesis—occurs during which the fatty acids are converted to acetic acid, carbon dioxide, and hydrogen
4. Methanogenesis—occurs during which the organic acids produced during the fermentation step are converted to methane and carbon dioxide

The efficiency of each phase is influenced by the temperature and the amount of time the process is allowed to react. For example, the organisms that perform hydrolysis and volatile acid fermentation (often called the acidogenic bacteria) are fast-growing microorganisms that prefer a slightly acidic environment and higher temperatures than the organisms that perform the methane formation step (the methanogenic bacteria).

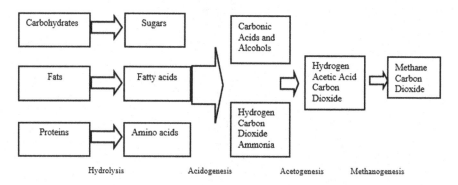

FIGURE 5.10   Key stages of anaerobic digestion.

TABLE 5.7
Typical Contents of Biogas

Matter	Percentage (%)
Methane, CH4	50–75
Carbon Dioxide, CO2	25–50
Nitrogen, N2	0–10
Hydrogen, H2	0–1
Hydrogen Sulfide, H2S	0–3
Oxygen, O2	0–2

A simplified generic chemical equation for the overall processes outlined above is as follows:

$$C_6H_{12}O_6 \rightarrow 3CO_2 + 3CH_4.$$

Biogas is the ultimate waste product of the bacteria feeding off the input biodegradable feedstock, and is mostly methane and carbon dioxide, with a small amount of hydrogen and trace hydrogen sulfide (see Table 5.7) (Spellman, 2009). Keep in mind that the ultimate output from a wastewater digester is water; biogas (methane) is more of an off-gas that can be used as an energy source. Wastewater digestion and the production of biogas are discussed in the next section.

ANAEROBIC DIGESTION OF SEWAGE BIOSOLIDS (SLUDGE)

Equipment used in anaerobic digestion typically includes a sealed digestion tank with either a fixed or a floating cover or an inflatable gas bubble, heating and mixing equipment, gas storage tanks, solids and supernatant withdrawal equipment, and safety equipment (e.g., vacuum relief, pressure relief, flame traps, explosion-proof electrical equipment). (**Caution**: Biosolids are inherently dangerous as possible sources of explosive gases and should never be entered without following OSHA's confined space entry permit requirements; only fully trained personnel should enter permit-required confined spaces).

In operation, process residual (thickened or unthickened sludge) is pumped into the sealed digester. The organic matter digests anaerobically by a two-stage process. Sugars, starches, and carbohydrates are converted to volatile acids, carbon dioxide, and hydrogen sulfide. The volatile acids are then converted to methane gas. This operation can occur in a single tank (single stage) or in two tanks (two stages). In a single-stage system, supernatant and/or digested solids must be removed whenever flow is added. In a two-stage operation, solids and liquids from the first stage flow into the second stage each time fresh solids are added. Supernatant is withdrawn from the second stage to provide additional treatment space. Periodically, solids are withdrawn for dewatering or disposal. The methane gas produced in the process may be used for many plant activities.

DID YOU KNOW?

The primary purpose of a secondary digester is to allow for solids separation.

Various performance factors affect the operation of the anaerobic digester. For example, % volatile matter in raw sludge, digester temperature, mixing, volatile acids/alkalinity ratio, feed rate, % solids in raw sludge, and pH are all important operational parameters that the operator must monitor.

Along with being able to recognize normal/abnormal anaerobic digester performance parameters, digester operators must also know and understand normal operating procedures. Normal operating procedures include sludge additions, supernatant withdrawal, sludge withdrawal, pH control, temperature control, mixing, and safety requirements. Important performance parameters are listed in Table 5.8 (Spellman 2009).

Sludge must be pumped (in small amounts) several times each day to achieve the desired organic loading and optimum performance.

Caution: Keep in mind that in fixed cover operations, additions must be balanced by withdrawals. If not, structural damage occurs.

Supernatant withdrawal must be controlled for maximum sludge retention time. When sampling, sample all draw-off points and select the level with the best quality. Digested sludge is withdrawn only when necessary—always leave at least 25% seed. pH should be adjusted to maintain 6.8 to 7.2 pH by adjusting feed rate, sludge withdrawal, or alkalinity additions.

Note: The buffer capacity of an anaerobic digester is indicated by the volatile acid/alkalinity relationship. Decreases in alkalinity cause a corresponding increase in ratio.

If the digester is heated, the temperature must be controlled to a normal temperature range of 90°–95° F. Never adjust the temperature by more than 1° F per day. If the digester is equipped with mixers, mixing should be accomplished to

TABLE 5.8
Anaerobic Digester—Sludge Parameters

Raw Sludge Solids	Impact
< 4% Solids	Loss of alkalinity
	Decreased sludge retention time
	Increased heating requirements
	Decreased Volatile Acid: Alk Ratio
4%–8% Solids	Normal operation
> 8% Solids	Poor mixing
	Organic overloading
	Decreased Volatile Acid: Alk Ratio

ensure organisms are exposed to food materials. Again, anaerobic digesters are inherently dangerous—several catastrophic failures have been recorded. To prevent such failures, safety equipment such as pressure relief and vacuum relief valves, flame traps, condensate traps, and gas collection safety devices are installed. It is important that these critical safety devices be checked and maintained for proper operation.

Note: As mentioned, because of the inherent danger involved with working inside anaerobic digesters, they are automatically classified as permit-required confined spaces. Therefore, all operations involving internal entry must be made in accordance with OSHA's confined space entry standard. If safety questions concerning safe entry into confined spaces (of any type) arise, consult with a Certified Safety Professional (CSP), Certified Industrial Hygienist (CIH), or Professional Engineer (PE).

During operation, anaerobic digesters must be monitored and tested to ensure proper operation. Testing should be accomplished to determine supernatant pH, volatile acids, alkalinity, BOD or COD, total solids, and temperature. Sludge (in and out) should be routinely tested for % solids and % volatile matter. Normal operating parameters are listed in Table 5.9 (Spellman, 2009).

Process control calculations involved with anaerobic digester operation include determining the required seed volume, volatile acid to alkalinity ratio, sludge retention time, estimated gas production, volatile matter reduction, and percent moisture reduction in digester sludge. Examples of how to make these calculations are provided in the following sections.

TABLE 5.9
Anaerobic Digester: Normal Operating Ranges

Parameter	Normal Range
Sludge Retention Time	
Heated	30–60 days
Unheated	180+ days
Volatile Solids Loading	0.04–0.1 lbs V.M/day/ft3
Operating Temperature	
Heated	90°–95° F
Unheated	Varies with season
Mixing	
Heated—primary	Yes
Unheated—secondary	No
% Methane in Gas	60%–72%
% Carbon Dioxide in Gas	28%–40%
pH	6.8%–7.2%
Volatile Acids: Alkalinity Ratio	≤ 0.1
Volatile Solids Reduction	40%–60%
Moisture Reduction	40%–60%

Required Seed Volume in Gallons

$$\text{Seed Volume (Gallons)} = (\text{Digester Volume}) (\% \text{ Seed}) \qquad (5.1)$$

EXAMPLE 5.1

Problem:

The new digester requires a 25% seed to achieve normal operation within the allotted time. If the digester volume is 266,000 gallons, how many gallons of seed material will be required?

Solution:

$$\text{Seed Volume} = 266,000 \text{ x } 0.25 = 66,500 \text{ gals}$$

Volatile Acids to Alkalinity Ratio

The *volatile acids to alkalinity ratio* can be used to control operation of an anaerobic digester.

$$\text{Ratio} = \frac{\text{Volatile Acids Concentration}}{\text{Alkalinity Concentration}} \qquad (5.2)$$

EXAMPLE 5.2

Problem:

The digester contains 240 mg/L volatile acids and 1,860-mg/L alkalinity. What is the volatile acids/alkalinity ratio?

$$\text{Ratio} = \frac{240 \text{ mg / L}}{1,860 \text{ mg / L}} = 0.13$$

Note: Increases in the ratio normally indicate a potential change in the operation condition of the digester, as shown in Table 5.10.

TABLE 5.10
Digestor Operating Conditions

Operating Condition	V.A./Alkalinity Ratio
Optimum	≤ 0.1
Acceptable range	0.–0.3
% Carbon dioxide in gas increases	≥ 0.5
pH decreases	≥ 0.8

SLUDGE RETENTION TIME

Sludge retention time is the length of time the sludge remains in the digester.

$$\text{SRT, Days} = \frac{\text{Digester Volume in Gallons}}{\text{Sludge Volume Added per Day, gpd}} \tag{5.3}$$

EXAMPLE 5.3

Problem:

Sludge is added to a 525,000-gallon digester at the rate of 12,250 gallons per day.

Solution:

$$\text{SRT} = \frac{525,000 \text{ gallons}}{12,250 \text{ gpd}} = 42.9 \text{ days}$$

ESTIMATED GAS PRODUCTION IN CUBIC FEET/DAY

The rate of gas production is normally expressed as the volume of gas (ft³) produced per pound of volatile matter destroyed. The total cubic feet of gas a digester will produce per day can be calculated by:

Gas Prod. (ft³) = Vol. Matter In, lbs/day x % Vol. Matter Red. x Prod. Rate ft³/lb

$$\tag{5.4}$$

EXAMPLE 5.5

Problem:

The digester receives 11,450 lbs of volatile matter per day. Currently, the volatile matter reduction achieved by the digester is 52%. The rate of gas production is 11.2 cubic feet of gas per pound of volatile matter destroyed.

Solution:

$$\text{Gas Prod.} = 11{,}450 \text{ lbs/day} \times 0.52 \times 11.2 \text{ ft}^3/\text{lb} = 66{,}685 \text{ ft}^3/\text{day}$$

VOLATILE MATTER REDUCTION, PERCENT

Because of the changes occurring during sludge digestion, the calculation used to determine percent volatile matter reduction is more complicated.

$$(\% \text{ Volatile Matter}_{in} - \% \text{ Volatile Matter}_{out}) \times 100$$

$$\% \text{ Reduction} = \frac{(\% \text{ Volatile Matter}_{in} - \% \text{ Volatile Matter}_{out}) \times 100}{\left[\% \text{ Volatile Matter}_{in} - (\% \text{ Volatile Matter}_{in} \times \% \text{ Volatile Matter}_{out}) \right]} \quad (5.5)$$

EXAMPLE 5.5

Problem:

Using the data provided below, determine the % Volatile Matter Reduction for the digester.

Raw Sludge Volatile Matter 74%

Digested Sludge Volatile Matter 55%

$$\% \text{Volatile Matter Reduction} = \frac{(0.74 - 0.55) \times 100}{\left[0.74 - (0.74 \times 0.55) \right]} = 57\%$$

PERCENT MOISTURE REDUCTION IN DIGESTED SLUDGE

$$\% \text{Moisture Reduction} = \frac{\left(\% \text{ Moisture}_{in} - \% \text{ Moisture}_{out}\right) \times 100}{\left[\% \text{ Moisture}_{in} - \left(\% \text{ Moisture}_{in} \times \% \text{ Moisture}_{out}\right) \right]} \quad (5.6)$$

EXAMPLE 5.6

Problem:

Using the digester data provided below, determine the % Moisture Reduction and % Volatile Matter Reduction for the digester.

Solution:

Raw Sludge % Solids	6%
Digested Sludge % Solids	14%

Note: % Moisture = 100% − Percent Solids

$$\%\text{Moisture Re duction} = \frac{(0.94 - 0.86)\ \text{X}100}{\left[0.94 - (0.94\ \text{X}0.86)\right]} = 61\%$$

ANAEROBIC DIGESTION OF AGRICULTURAL WASTES

Animal waste accounts for 10% of methane emissions in the United States. When not correctly managed, farm waste slurries can also seriously pollute local water courses. Small anaerobic digesters have been installed on farms to treat excess animal slurries, which can't be placed on the land. The biogas formed is normally used for heat but can also be used to fuel engines and other on-site energy needs such as electricity and heating. On-site biogas production and management also reduce offensive odors from overloaded or improperly managed manure storage facilities. These odors impair air quality and may be a nuisance to nearby communities. Anaerobic digestion of animal waste reduces these offensive odors because the volatile organic acids, the odor-causing compounds, are consumed by biogas-producing bacteria. In addition to biogas, another important by-product of anaerobic digestion is ammonium, which is the major constituent of commercial fertilizer, which is readily available and utilized by crops. The bottom line on the production of biogas on the farm: Biogas recovery can improve profitability while improving environmental quality.

LANDFILL BIOGAS

Landfills can be a source of energy. Landfills produce methane as organic waste decomposes in the same anaerobic digestion process used in converting wastewater and farm waste slurries into biogas. Most landfill gas results from the degradation of cellulose contained in municipal and industrial solid waste. Unlike animal manure digesters, which control the anaerobic digestion process, the digestion occurring in landfills is an uncontrolled process of biomass decay. To be technically feasible, a landfill must be at least 40 feet deep and have at least one million tons of waste in place for landfill gas collection.

Landfill Biogas

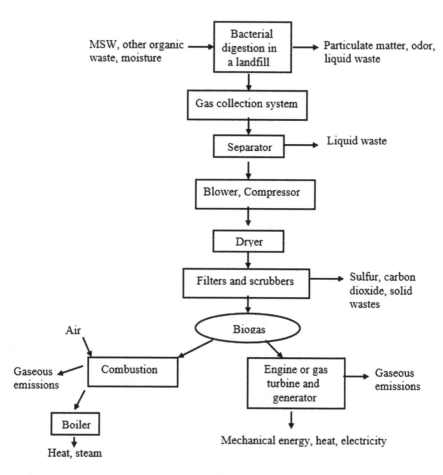

FIGURE 5.11   Landfill biogas system flow diagram.

The efficiency of the process depends on the waste composition and moisture content of the landfill, cover material, temperature, and other factors. The biogas released from landfills, commonly called "landfill gas," is typically 50% methane, 45% carbon dioxide, and 5% other gases. The energy content of landfill gas is 400 to 550 Btu per cubic foot.

Figure 5.11 shows a landfill energy system. Such a system consists of a series of wells drilled into the landfill. A piping system connects the wells and collects the gas. Dryers remove moisture from the gas and filters remove impurities. The gas typically fuels an engine-generator set or gas turbine to produce electricity. The gas can also fuel a boiler to produce heat or steam. Because waste-generated biogas is a "dirty gas," as compared to natural gas, further gas cleanup is required to improve biogas to pipeline quality, the equivalent of natural gas. Reforming the gas to hydrogen would make possible the production of electricity using fuel cell technology.

BIOMASS FOR BIOPOWER

In addition to advanced fuels, biomass can also be used to produce biopower. This can be done in several ways, including direct combustion of biomass in dedicated power plants, co-firing biomass with coal, biomass gasification in a combined cycle plant to produce steam and electricity, or via anaerobic digestion (EPRI, 1997).

Combustion is the burning of biomass in air. This involves the conversion of chemical energy stored in biomass into heat, mechanical power, or electricity (McKendry, 2002). While it is possible to use all types of biomass, combustion is preferable when biomass is more than 50% dry. High-moisture biomass is better suited for biological conversion processes. Net bioenergy conversion efficiencies for biomass combustion power plants range from 20% to 40%. Higher efficiencies are obtained with combined heat and power (CHP) facilities and with large-size power-only systems (over 100 mega-watt-electrical—Mwe), or when the biomass is co-fired with coal in power plants (McKendry, 2002).

Co-firing biomass with coal is a straightforward and inexpensive way to diversify the fuel supply, reduce coal plant air emissions (NO_x, SO_2, CO_2), divert biomass from landfills, and stimulate the biomass power industry (Hughes, 2000). Moreover, biomass is the only renewable energy technology that can directly displace coal. Given the dominance of coal-based power plants in US electricity production, co-firing with biomass fuel is the most economical way to reduce greenhouse gas emissions. Possible biomass fuel for co-firing includes wood waste, short-rotation woody crops, switchgrass, alfalfa stems, various types of manure, landfill gas, and wastewater treatment gas (Tillman, 2000). In addition, agricultural residues such as straw can also be used for co-firing.

A promising technology development currently at the demonstration stage is the biomass integrated gasification/combined cycle (BIG/CC), where a gas turbine converts the gaseous fuel to electricity with a high conversion efficiency, reaching 40% to 50% of the heating value of the incoming gas (McKendry, 2002). An important advantage of gasification is the ability to work with a wider variety of feedstocks, such as high-alkali fuels that are problematic with direct combustion. High-alkali fuels such as switchgrass, straw, and other agricultural residues often cause corrosion, but the gasification systems can easily remove the alkali species from the fuel gas before it is combusted. High silica, also a problem with grasses, can result in slagging in the reactor.

National Renewable Energy Laboratory (NREL, 1993) points out that the slagging problem is not unique to one form of biomass but instead is common with many different types of biomass fuels. Slagging deposits can reduce heat transfer, reduce combustion efficiency, and damage combustion chambers when large particles break off. Research has focused on two alkali metals, potassium and sodium, and silica, all elements commonly found in living plants. In general, it appears that faster growing plants (or faster-growing plant components such as seeds) tend to have higher concentrations of alkali metal and silica. Thus, material such as straw, nut hulls, fruit pits, weeds, and grasses tend to create more problems when burned than wood from a slow-growing tree.

Potassium and sodium metals, whether in the form of oxides, hydroxides, or metallo-organic compounds, tend to lower the melting point of ash mixtures containing various other minerals such as silica ($SiO2$). The high alkali content (up to 35%) in the ash from burning annual crop residues lowers the fusion, or "sticky temperature," of these ashes from 2200° F for wood ash to as low as 1300° F. This results in serious slagging on the boiler grate or in the bed, and fouling of convection heat transfer surfaces. Even small percentages (10%) of some of these high-alkali residues burned with wood in conventional boilers will cause serious slagging and fouling in a day to two, necessitating combustion system shutdown.

A method to predict slagging and fouling from combustion of biomass fuels has been adapted from the coal industry. The method involves calculating the weight in pounds of alkali ($K20 + Na2O$) per million Btu in the fuel as follows:

$$\frac{1 \times 10^6}{Btu\,/\,lb} \times \%\,Ash \times \%\,Alkali\,of\,the\,Ash = \frac{lb\;Alkali}{MM\;Btu}.$$

This method combines all the pertinent data into one index number. A value below 0.4 lb/MM Btu (MM Btu = million Btu) is considered a low slagging risk. Values between 0.4 and 0.8 lb/MM Btu will probably slag, with increasing certainty of slagging as 0.8 lb/MM Btu is approached. Above 0.8 lb/MM Btu, the fuel is virtually certain to slag and foul (see Table 5.11).

As mentioned, another process for biomass is the application of anaerobic digestion to produce biogas (methane) for electricity generation. Recall that anaerobic

---

TABLE 5.11
Alkali Content and Slagging Potential of Various Biofuels

Fuel	Total Alkali (lb/MM Btu)	Slagging Potential
WOOD		
Pine Chips	0.07	Minimal
White Oak	0.14	Minimal
Hybrid Poplar	0.46	Probable
Urban Wood Waste	0.46	Probable
Clean Tree Trimmings	0.73	Probable
PITS, NUTS, SHELLS		
Almond Shells	0.97	Certain
Refuse-derived Fuel	1.60	Certain
GRASSES		
Switchgrass	1.97	Certain
Wheat Straw—average	2.00	Certain
Wheat Straw—hi alkali	5.59	Certain
Rice Straw	3.80	Certain

Source: Adaptation from Miles et al. (1993).

digestion involves the controlled breakdown of organic wastes by bacteria in the absence of oxygen. Major agricultural feedstocks for anaerobic digestion include food processing wastes and manure from livestock operations. The Energy Information Agency also projects a significant increase in generation of electricity from municipal waste and landfill gas—to about 0.5% of US electricity consumption.

BIOMASS FOR BIOPRODUCTS[4]

Bioproducts are industrial and consumer goods manufactured wholly or in part from renewable biomass (plant-based resources). Today's industrial bioproducts are amazingly diverse, ranging from solvents and paints to pharmaceuticals, soaps, cosmetics, and building materials (see Table 5.12). Industrial bioproducts are integral to our way of life—few sectors of the economy do not rely in some way on products made from biomass.

Corn, wood, soybeans, and plant oils are the primary resources used to create this remarkable diversity of industrial and consumer goods. In some cases, it is not readily apparent that a product is derived in part from biomass. Biomass components are often combined with other materials such as petrochemicals and minerals to manufacture the final product. Soybean oil, for example, is blended with other components to produce paints, toiletries, solvents, inks, and pharmaceuticals. Some products, such as starch adhesives, are derived entirely from biomass.

The many derivatives of corn illustrate the diversity of products that can be obtained from a single biomass resource. Besides being an important source of food and feed, corn serves as a feedstock for ethanol and sorbitol (a sweetish, crystalline alcohol), industrial starches and sweeteners, citric and lactic acid, and many other products.

Biomass, which is comprised of carbohydrates, can be used to produce some of the products that are commonly manufactured from petroleum and natural gas, or hydrocarbons. Both resources contain the essential elements of carbon and hydrogen. In some cases, both resources have captured a portion of market share (see Table 5.13).

TABLE 5.12
Common Products from Biomass

Biomass Resource	Uses
Corn	Solvents, pharmaceuticals, adhesives, starch, resins, binders, polymers cleaners, ethanol.
Vegetable Oils	Surfactants in soaps and detergents, pharmaceuticals (inactive ingredients), inks, paints, resins, cosmetics, fatty acids, lubricants, biodiesel
Wood	Paper, building materials, cellulose for fibers and polymers, resins, binders, adhesives, coatings, paints, fatty acids, road and roofing pitch

Source: USDOE (2003a).

TABLE 5.13
Products from Hydrocarbons versus Carbohydrates

Product	Total Production (millions of tons)	% Derived from Plants
Adhesives	5.0	40
Fatty Acids	2.5	40
Surfactants	3.5	35
Acetic Acid	2.3	17.5
Plasticizers	0.8	15
Activated Carbon	1.5	12
Detergents	12.6	11
Pigments	15.5	6
Dyes	4.5	6
Wall Points	7.8	3.5
Inks	3.5	3.5
Plastics	30	1.8

Source: ILSR (2002).

CLASSES OF BIOPRODUCTS

The thousands of different industrial bioproducts produced today can be categorized into five major areas:

- *Sugar and starch bioproducts* derived through fermentation and thermochemical processes include alcohols, acids, starch, xanthum gum, and other products derived from biomass sugars. Primary feedstocks include sugarcane, sugar beets, corn, wheat, rice, potatoes, barley, sorghum grain, and wood.
- *Oil- and lipid-based bioproducts* include fatty acids, oils, alkyd resins, glycerin, and a variety of vegetable oils derived from soybeans, rapeseed, or other oilseeds.
- *Gum and wood chemicals* include tall oil (liquid rosin), alkyd resins, rosins, pitch, fatty acids, turpentine, and other chemicals derived from trees.
- *Cellulose derivatives, fibers, and plastics* include products derived from cellulose, including cellulose acetate (cellophane) and triacetate, cellulose nitrate, alkali cellulose, and regenerated cellulose. The primary sources of cellulose are bleached wood pulp and cotton linters.
- *Industrial enzymes* are used as biocatalysts for a variety of biochemical reactions in the production of starch and sugar, alcohols, and oils. They are also used in laundry detergents, tanning of leathers, and textile sizing (Uhlig 1998).

SIDEBAR 5.2 THE BRAZILIAN EXAMPLE[5]

Brazil has the world's second-largest ethanol program and is capitalizing on plentiful soybean supplies to expand into biodiesel. More than half of the nation's sugarcane crop is processed into ethanol, which now accounts for about 20% of the country's fuel supply.

Initiated in the 1970s after the Organization of the Petroleum Exporting Countries oil embargo, Brazil's policy program was designed to promote the nation's energy independence and to create an alternative and value-added market for sugar producers. The government has spent billions to support sugarcane producers, develop distilleries, build up a distribution infrastructure, and promote production of pure-ethanol-burning and, later, flex-fuel vehicles (able to run on gasoline, ethanol-gasoline blends, or pure hydrous (or wet; 7% water content) ethanol). Advocates contend that, while the costs were high, the program saved far more in foreign exchange from reduced petroleum imports.

In the mid- to late 1990s, Brazil eliminated direct subsidies and price setting for ethanol. It pursued a less intrusive approach with two main elements—a blending requirement (now about 25%) and tax incentives favoring ethanol use and the purchase of ethanol-using or flex-fuel vehicles. Today, more than 80% of Brazil's newly produced automobiles have flexible fuel capability, up from 30% in 2004. With ethanol widely available at almost all of Brazil's 32,000 gas stations, Brazilian consumers currently choose primarily between 100% hydrous ethanol and a 2% ethanol-gasoline blend based on relative prices.

Approximately 20% of current fuel use (alcohol, gasoline, and diesel) in Brazil is ethanol, but it may be difficult to raise the share as Brazil's fuel demand grows. Brazil is a middle-income economy with per capita energy consumption only 15% that of the United States and Canada. Current ethanol production levels in Brazil are not much higher than they were in the late 1990s. Production of domestic off- and onshore petroleum resources has grown more rapidly than ethanol and accounts for a larger share of expanding fuel use than ethanol in the last decade.

THE BOTTOM LINE

Biomass energy is becoming increasingly popular as a renewable and sustainable energy source.

With near-record oil prices, the future of biofuel—made from plant material (biomass)—is of keen interest worldwide. Using biofuels to power industry, private vehicles, and personal appliances has several benefits over the use of conventional fuels. For example, environmental benefits include the use of biomass energy to greatly reduce greenhouse gas emissions. Burning biomass releases about the same amount of carbon dioxide as burning fossil fuels. However, fossil fuels release carbon dioxide captured by photosynthesis millions of years ago—an essentially "new" greenhouse gas. Biomass, on the other hand, releases carbon dioxide that is largely balanced by the carbon dioxide captured in its own growth (depending on how much energy was used to grow, harvest, and process the fuel).

Another benefit of biomass use for fuel is that it can reduce dependence on foreign oil because biofuels are the only renewable liquid transportation fuels available.

Moreover, biomass energy supports US agricultural and forest product industries. The main biomass feedstocks for power are paper mill residue, lumber mill scrap, and municipal waste. For biomass fuels, the feedstocks are corn (for ethanol) and soybeans (for biodiesel), both surplus crops. Soon—and with developed technology— agricultural residues such as corn stover (the stalks, leaves, and husks of the plant) and wheat straw will also be used. Long-term plans include growing and using dedicated energy crops, such as fast-growing trees and grasses that can grow sustainably on land that will not support intensive food crops.

The preceding lists many of the benefits of using biomass fuel. That is all well and good, but the reality is that the future role of biofuels depends on profitability and new technologies. Technological advances and efficiency gains—higher biomass yields per acre and more gallons of biofuel per ton of biomass—could steadily reduce the economic cost and environmental impact of biofuel production. Biofuel production will likely be most profitable and environmentally benign in tropical areas where growing seasons are longer, per acre biofuel yields are higher, and fuel and other input costs are lower. For example, Brazil uses bagasse, which is a by-product from sugar production, to power ethanol distilleries, whereas the United States uses natural gas or coal.

Biofuels will most likely be part of a portfolio of solutions to high oil prices, including conservation and the use of other alternative fuels. The role of biofuels in global fuel supplies is likely to remain modest because of its land intensity. In the United States, replacing all current gasoline consumption with ethanol would require more land in corn production than is presently in all agricultural production. Technology will be central to boosting the role of biofuels. If the energy of widely available cellulose materials could be economically harnessed around the world, biofuel yards per acre could more than double, reducing land requirements significantly (USDA, 2007).

NOTES

1   This section is adapted from F.R. Spellman (2009) *Biology for the Non-Biologist*. Lanham, MD: Government Institutes Press.
2   Much of this section is from USDA et al. (2008) *The Economics of Biomass Feedstocks in the United States*. Washington, DC: Biomass Research and Development Board.
3   Information from USDA, *Amber Waves* Feb. 2008 edition. Ephraim Leibtag. Accessed 03/ 08/10 @ www.ers.usda.gov/amberwaves/February08/Featrues/CornPrices.htm
4   Information in this section is from USDOE (2003a) *Industrial Bioproducts: Today and Tomorrow*. Washington, DC: Office of the Biomass Program.
5   Example is taken from USDA (2007): *Amber Waves: The Future of Biofuels; a Global Perspective*. November.

REFERENCES AND RECOMMENDED READING

Aylott, M.J. 2008. Yield and spatial supply of bioenergy poplar and willow short-rotation coppice in the UK. *New Phytologist* **178**(2): 358–370.
Baize, J. 2006. Bioenergy and Biofuels. Agricultural Outlook Forum, Feb. 17. Washington, DC: US Department of Agriculture.
Biodiesel. 2007. FUMPA Biofuels. Biodiesel Magazine, Feb.
Christi, Y. 2007. Biodiesel from microalgae. *Biotechnology Advances* **25**: 294–306.

Efficiency & Renewable Energy (EERE). 2008. *Biomass Program.* Accessed 03/04/20 @ www1.eere.energy.gov/biomass/feedstocks_types.html

Electric Power Research Institute (EPRI). 1997. Renewable Energy Technology Characterizations. Dec. Washington, DC: TR-109496, US Department of Energy.

Fahey, J. 2001. Shucking petroleum. Forbes Magazine, Nov. 26, p. 206.

Friedman, T.L. 2010. The fat lady has sung. New York Times 02/20/10.

Graham, R., Nelson, R., Sheehan J., Perlack, R. and Wright L. 2007. Current and potential US corn stover supplies. *Agronomy Journal* **99**: 1–11.

Hughes, E. 2000. Biomass cofiring: Economics, policy and opportunities. *Biomass and Bioenergy* **19**: 45–65.

Institute for Local Self-Reliance (ILSR). 2002. *Accelerating the Shift to a Carbohydrate Economy: The Federal Role: Executive Summary of the Minority Report of the Biomass Research & Development Technical Advisory Committee.* Washington, DC: Institute for Local Self-Reliance.

McDill, M. 2010, 2009. Symposium on system analysis. Mathematical and Computational Forestry and Natural Resources **2**(2): 97–98.

McKeever, D. 2004. *Inventories of Woody Residues and Solid Wood Waste in the United States. 2002 Ninth International Conference, Inorganic-Bonded Composite Materials.* Oct. 10–13, Vancouver, BC.

McKendry, P. 2002. Energy production from biomass (part I): Overview of biomass. *Bioresource Technology* **83**: 37–46.

McLaughlin, S.B. and Kzos, L.A. 2005. Development of switchgrass as a bioenergy feedstock in the U.S. *Biomass & Bioenergy* **28**: 515–535.

Miles, T.R., Miles, T.R., Jr., Baxter, L.L., Jenkins, B.M., and Oden, L.L. 1993. *Alkali Slagging Problems with Biomass Fuels. First Biomass Conference of the American Energy, Environment, Agriculture, and Industry*, Volume 1. Burlington, VT.

National Institute of Standards and Technology (NIST). 1994. *Handbook 44 Appendix D Definitions.* Washington, DC: US Department of Commerce.

National Renewable Energy Laboratory (NREL). 1993. *Alkali Content and Slagging Potential of Various Biofuels.* Accessed 03/17/20 @ http://cta.ornl.gov/bedb/biopower/alkali_conten_salgging_potential_of_various biofules.xls

Parrish, D.J., Fike, J.H., Bransby, D.I., and Samson, R. 2008. Establishing and managing switchgrass as an energy crop. *Forage and Grazinglands.* DOI:10.1094/FG-2008-0220-01-RV

Polagye, B., Hodgson, K., and Malte, P. 2007. An economic analysis of bioenergy options using thinning from overstocked forests. *Biomass and Bioenergy* **31**: 105–125.

Rinehart, L. 2006. *Switchgrass as a Bioenergy.* Butte, MT: National Center for Appropriate Technology.

Samson, R. et al. 2008. Developing energy crops for thermal applications: Optimizing fuel quality, energy security and GHG mitigation. In *Biofuels, Solar and Wind as Renewable Energy Systems: Benefits and Risks.* D. Pimental (Ed.). Berlin: Springer Science.

Schenk, P., Thomas-Hall, S., Stephens, R., Marx U., Mussgnug, J., Posten, C., Kruse, O., Hankamer, B. 2008. Second generation biofuels: High-efficiency microalgae for industrial production. *BioEnergy Research* **1**(1): 20–45.

Spellman, F.R. 2009. *Handbook of Water and Wastewater Treatment Plant Operations*, 2nd edition. Boca Raton, FL: CRC Press.

Taipei Times. 2008. Algae eyed as biofuel alternative. January 12. Accessed 7/27/20 @ www.taipietimes.com/news/taiwan/archives/2008/01-12-2003396760

Tillman, D. 2000. Biomass cofiring: The technology, the experience, the combustion consequences. *Biomass and Bioenergy* **19**: 365–384.

Time online. 2005. Seaweed to breathe new life into fight against global warming. Accessed 7/ 26/10 @ www.timeonline.Co.Uk/tot/news.world/article522203.ece

Uhlig, H. 1998. *Industrial Enzymes and Their Applications*. New York: John Wiley & Sons.

US Department of Agriculture (USDA). 2007. *Amber Waves: Future of Biofuels*. Accessed 03/ 20/21 @ www.ers.usda.gov/AmberWaves/november07/features/biofuels.htm

US Department of Agriculture (USDA)-World Agricultural Outlook Board. 2008. *World Agricultural Supply and Demand Estimates*. Washington, DC: US Department of Agriculture.

US Department of Energy (USDOE). 2003a. *Industrial Bioproducts: Today and Tomorrow*. Washington, DC: US Department of Energy.

US Department of Energy (USDOE). 2003b. *The Bioproducts Industry: Today and Tomorrow*. Washington, DC: US Department of Energy.

US Department of Energy (USDOE). 2007. *Understanding Biomass: Plant Cell Walls*. Washington, DC: US Department of Energy.

US Department of Energy (USDOE). 2010. *Fuel Ethanol Production*. Accessed 7/25/22 @ http://genomicScience.energy.gov/biofuels.ethanolproduction.shtml

US Environmental Protection Agency (USEPA). 1979. *Process Design Manual Sludge Treatment and Disposal*. Washington, DC: US Environmental Protection Agency, Office of Research and Development, EPA 625/625/1-79-011.

US Environmental Protection Agency (USEPA). 2006. *Biosolids Technology Fact Sheet: Multi-Stage Anaerobic Digestion*. Washington, DC: US Environmental Protection Agency, Office of Water, EPA832-F-06-031.

US Environmental Protection Agency (USEPA). 2007. *Fuel Economy Impact Analysis of RFG*. Accessed 03/11/23 @ www.epa.gov/oms/rfsecon.htm

Wiltsee, G., 1998. *Urban Wood Waste Resource Assessment NREL/SR-570-25918*. Golden, CO: National Renewable Energy Laboratory.

6 Geothermal Energy

INTRODUCTION[1]

Approximately 4,000 miles below the Earth's surface is the core, where temperatures can reach 9000° F. This heat—geothermal energy (*geo*, meaning earth, and *thermos*, meaning hea*t*)—flows outward from the core, heating the surrounding area, which can form underground reservoirs of hot water and steam. These reservoirs can be tapped for a variety of uses, such as to generate electricity or heat buildings.

In 2008, scientists with the US Geological Survey (USGS) completed an assessment of our nation's geothermal resources. Geothermal power plants are currently operating in six states: Alaska, California, Hawaii, Idaho, Nevada, and Utah. The assessment points out that the electric power generation potential from identified geothermal system is 9,057 Megawatts-electric (MWe), distributed in 13 states. The mean estimated power production potential from undiscovered geothermal resources is 30.033 MWe. Moreover, another estimated 517,800 MWe could be generated through implementation of technology for creating geothermal reservoirs in regions characterized by high temperature, but low-permeability rock formations (USGS 2008).

The geothermal energy potential in the uppermost 6 miles of the Earth's crust amounts to 50,000 times the energy of all oil and gas resources in the world. In the United States, most geothermal reservoirs are in the western states, Alaska, and Hawaii. However, geothermal heat pumps (GHPs), which take advantage of the shallow ground's stable temperature for heating and cooling buildings, can be used almost anywhere.

DID YOU KNOW?

Scientists estimate that geothermal potential could be as large as 100 million kW (Kutsher, 2000).

Again, it is important to point out that there is nothing new about renewable energy. From solar power to burning biomass (wood) in the cave and elsewhere, humans have taken advantage of renewable resources since time immemorial. For example, hot springs have been used for bathing since Paleolithic times or earlier (USDOE, 2009). The early Romans used hot springs to feed public baths and for underfloor heating. The world's oldest geothermal district heating system, in France, has been operating since the fourteenth century (Lund, 2007). The history of geothermal energy use in the United States is interesting and lengthy. In the following a brief chronology of major geothermal events in the United States is provided (USDOE, 2006).

DID YOU KNOW?

The US Geological Survey has calculated the heat energy in the upper 10 kilometers of the earth's crust in the U.S. is equal to over 600,000 times the country's annual non-transportation energy consumption. Probably no more than a tiny fraction of this energy could ever be extracted economically. However, just one hundredth of 1% of the total is equal to half the country's current non-transportation energy needs for more than a century, with only a fraction of the pollution from fossil-fueled energy sources.

–McLarty et al. (2000, p.3793)

GEOTHERMAL TIME LINE

8,000 BCE (and earlier)

Paleo-Indians used hot springs for cooking, and for refuge and respite. Hot springs were neutral zones where members of warring nations would bathe together in peace. Native Americans have a history with every major hot spring in the United States.

1807

As European settlers moved westward across the continent, they gravitated toward these springs of warmth and vitality. In 1807, the first European to visit the Yellowstone area, John Colter (c. 1774–c. 1813), widely considered to be the first mountain man, probably encountered hot springs, leading to the designation "Colter's Hell." Also in 1897, settlers founded the city of Hot Springs, Arkansas, where, in 1830, Asa Thompson charged one dollar each for the use of three spring-fed baths in a wooden tub, and the first known commercial use of geothermal energy occurred.

1847

William Bell Elliot, a member of John C. Fremont's survey party, stumbles upon a steaming valley just north of what is now San Francisco, California. Elliot calls the area The Geysers—a misnomer—and thinks he has found the gates of Hell.

1852

The Geysers is developed into a spa called The Geysers Resort Hotel. Guests include J. Pierpont Morgan, Ulysses S. Grant, Theodore Roosevelt, and Mark Twain.

1862

At springs located southeast of The Geysers, businessman Sam Brannan pours an estimated half million dollars into an extravagant development dubbed "Calistoga," replete with hotel, bathhouse, skating pavilion, and racetrack. Brannan's was one of many spas reminiscent of those of Europe.

1864

Homes and dwellings have been built near springs through the millennia to take advantage of the natural heat of these geothermal springs, but the construction of the Hot Lake Hotel near La Grande, Oregon, marks the first time that the energy from hot springs is used on a large scale.

1892

Boise, Idaho, provides the world's first district heating system as water is piped from hot springs to town buildings. Within a few years, the system will serve 200 homes and 40 downtown businesses. Today, there are four district heating systems in Boise that provide heat to over 5 million square feet of residential, business, and governmental space. There are now 17 district heating systems in the United States and dozens more around the world.

1900

Hot springs water is piped to homes in Klamath Falls, Oregon.

1921

John D. Grant drills a well at The Geysers with the intention of generating electricity. This effort is unsuccessful, but one year later Grant meets with success across the valley at another site, and the United States' first geothermal power plant goes into operation. Grant uses steam from the first well to build a second well, and several wells later, the operation is producing 250 kilowatts, enough electricity to light the buildings and streets at the resort. The plant, however, is not competitive with other sources of power, and it soon falls into disuse.

1927

Pioneer Development Company drills the first exploratory well at Imperial Valley, California.

1930

The first commercial greenhouse use of geothermal energy is undertaken in Boise, Idaho. The operation uses a 1,000-foot well drilled in 1926. In Klamath Falls, Charlie Lieb develops the first downhole heat exchanger (DHE) to heat his house. Today, more than 500 DHEs are in use around the country.

1940

The first residential space heating in Nevada begins in the Moan area in Reno.

1948

Geothermal technology moves east when Carl Nielsen develops the first ground-source heat pump, for use at his residence. J.D. Krocker, an engineer in Portland, Oregon, pioneered the first commercial building use of a groundwater heat pump.

1960

The country's first large-scale geothermal electricity-generating plant begins operation. Pacific Gas and Electric operates the plant located at the Geysers. The first turbine produces 11 megawatts (MW) of net power and has operated successfully for more than 30 years. Today, 69 generating facilities are in operation at 18 resource sites around the country.

1978

Geothermal Food Processors, Inc. opens the first geothermal food-processing (crop-drying) plant in Brady Hot Springs, Nevada. The Load Guaranty Program provides $3.5 million for the facility.

1979

The first electrical development of a water-dominated geothermal resource occurs, at the east Mesa field in the Imperial Valley in California. The plant is named for B.C. McCabe, the geothermal pioneer who, with his Magma Power Company, did field development work at several sites, including The Geysers.

1980

TAD's Enterprises of Nevada pioneers the use of geothermal energy for the cooking, distilling, and drying processes associated with alcohol fuels production. UNOCAL builds the country's first flash plant, generating 10 MW at Brawley, California.

1982

Economical electrical generation begins at California's Salton Sea geothermal field using crystallizer-clarifier technology. The technology resulted from a government/industry effort to manage the high-salinity brines at the site.

1984

A 20-MW plant begins generating power at Utah's Roosevelt Hot Springs. Nevada's first geothermal electricity is generated with a 1.3-MW binary power plant that begins operation.

1987

Geothermal fluids are used in the first geothermal-enhanced heap leaching project for gold recovery, near Round Mountain, Nevada.

1989

The world's first hybrid (organic Rankine/gas engine) geopressure-geothermal power plant begins operation at Pleasant Bayou, Texas, using both the heat and the methane of a geopressured resource.

1992

Electrical generation begins at the 25-MW geothermal plant in the Puna field of Hawaii.

1993

A 23-MW binary power plant is completed at Steamboat Springs, Nevada.

1995

Integrated Ingredients dedicates a food-dehydration facility that processes 15 million pounds of dried onions and garlic per year at Empire, Nevada. A USDOE low-temperature resource assessment of 10 western states identifies nearly 9,000 thermal wells and springs and 271 communities collocated with a geothermal resource greater than 50.

2002

Organized by GeoPowering the West, geothermal development working groups are active in five states—Nevada, Idaho, New Mexico, Oregon, and Washington. Group members represent all stakeholder organizations. The working groups are identifying barriers to geothermal development in their state and bringing together all interested parties to arrive at mutually beneficial solutions.

2003

The Utah Geothermal Working Group is formed.

Table 6.1 highlights geothermal energy's quadrillion Btu ranking in current renewable energy source use. As pointed out in the table, the energy consumption by energy source computations are from the year 2007. Thus, the 0.353 geothermal quadrillion Btu figure is expected to steadily increase to an increasingly higher level and this should be reflected in the 2022–2023 figures when they are released.

GEOTHERMAL ENERGY: THE BASICS

As mentioned, geothermal energy processes evolve around the natural heat of the earth that can be used for beneficial purposes when the heat is collected and transported to the surface. To gain proper understanding of geothermal energy as it is used at present, it is important to define enthalpy, that is, the heat content of a substance per unit mass. The term "enthalpy" is used because temperature alone is not sufficient to define the useful energy content of a steam/water mixture. A mass of steam at a given temperature and pressure can provide much more energy than the same mass of water under the same conditions. Enthalpy is a function of:

$$Pressure + Volume + Temperature = Enthalpy$$

TABLE 6.1
US Energy Consumption by Energy Source, 2022 (%)

Energy Source	2022
Total	**97.33**
Renewable	12.16
Biomass (biofuels, waste, wood and wood-derived)	40%
Biofuels	19%
Waste	4%
Wood-derived Fuels	17%
Geothermal	**2%**
Hydroelectric Conventional	19%
Solar/PV	12%
Wind	27%

Source: EIA 2022. *US Energy consumption by Energy Source*. Accessed 06/25/23 @ www.eia.energyex plained/us-energy-facts.

Geothermal practitioners usually classify geothermal resources as "high enthalpy" (water and steam at temperatures above about 180°–200° C), "medium enthalpy" (about 100°–180° C), and "low enthalpy" (< 100° C). For the present text, it is sufficient to think of temperature and enthalpy as going hand in hand.

EARTH'S LAYERS

Earth is made up of three main compositional layers: crust, mantle, and core. The crust has variable thickness and composition: Continental crust is 10–50 km thick, while the oceanic crust is 8–10 km thick. The elements silicon, oxygen, aluminum, and ion make up the Earth's crust. Like the shell of an egg, the Earth's crust is brittle and can break. Earth's continental crust is 35 kilometers thick. The oceanic crust is 7 kilometers thick.

Based on seismic (earthquake waves) that pass through the Earth, we know that below the crust is the mantle, a dense, hot layer of semi-solid (plastic-like liquid) rock approximately 2,900 km thick. The mantle, which contains silicon, oxygen, aluminum, and more iron, magnesium, and calcium than the crust, is hotter and denser because temperature and pressure inside the Earth increase with depth. According to the USGS (2008), as a comparison, the mantle might be thought of as the white of a boiled egg. The 30 km-thick transitional layer between the mantle and the crust is called the Moho layer. The temperature at the top of the mantle is 870° C. The temperature at the bottom of the mantle is 2,200° C.

At the center of the Earth lies the core, which is nearly twice as dense as the mantle because its composition is metallic (Iron (Fe)-Nickel (NI) alloy) rather than stony. Unlike the yolk of an egg, however, the Earth's core is made up of two distinct parts: a 2,200 km-thick liquid outer core and a 1,250 km-thick solid inner core. As the Earth rotates, the liquid outer core spins, creating the Earth's magnetic field. Several important points related to Earth's structure and geothermal properties include the following:

- Heat flows outward from the center because of radioactive decay.
- The crust (about 30 and 60 km thick), insulates us from the interior heat.
- A solid inner core followed by a liquid outer core, with a semi-molten mantle.
- Temperature at the base of the crust is about 1000° C, increasing slowly into the core.
- Hot spots are located 2 to 3 km from the surface.

CRUSTAL PLATES

Within the past 45 to 50 years, geologists have developed the theory of plate tectonics (tectonics: Greek, "builder"). The theory of plate tectonics deals with the formation, destruction, and large-scale motions of great segments of Earth's surface (crust), called *plates*. This theory relies heavily on the older concepts of continental drift (developed during the first half of the twentieth century), and seafloor spreading (understood during the 1960s), which help explain the cause of earthquakes and volcanic eruptions, and the origin of fold mountain systems.

Regarding crustal plate activity and geothermal energy, and as mentioned, large quantities of heat (that are economically extractable) tend to be concentrated in places where hot or even molten (magma) exists at relatively shallow depths in the Earth's outmost layer (the crust). Such "hot" zones are generally near the boundaries of the dozen or so slabs of rigid rock (or plates) that form the Earth's lithosphere.

These crustal plates are composed of great slabs of rock (lithosphere), about 100 km thick, and cover many thousands of square miles (they are thin in comparison to their length and width); they float on the ductile asthenosphere, carrying both continents and oceans. Many geologists recognize at least eight main plates and numerous smaller ones. These *main* plates include

- African Plate covering Africa—Continental plate
- Antarctic Plate covering Australia—Continental plate
- Australian Plate covering Australia—Continental plate
- Eurasian Plate covering Asia and Europe—Continental plate
- Indian Plate covering Indian subcontinent and a part of Indian Ocean—Continental plate
- Pacific Plate covering the Pacific Ocean—Oceanic plate
- North American Plate covering North America and northeast Siberia—Continental plate
- South American Plate covering South America—Continental plate

The *minor* plates include

- Arabian Plate
- Caribbean Plate
- Juan de Fuca Plate

- Cocos Plate
- Nazea Plate
- Philippine Plate
- Scotia Plate

The plates literally ride in the asthenosphere, which is the ductile, soft, plastic-like zone in the upper mantle. Crustal plates move in relation to one another at one of three types of plate boundaries: convergent (collision boundaries), divergent (spreading boundaries), and transform. These boundaries between plates are typically associated with deep-sea trenches, large faults, fold mountain ranges, and mid-oceanic ridges.

CONVERGENT BOUNDARIES

Convergent boundaries (or active margins) develop where two plates slide towards each other, commonly forming either a subduction zone (if one plate subducts or moves underneath the other), or a continental collision (if the two plates contain continental crust). To relieve the stress created by the colliding plates, one plate is deformed and slips below the other.

DIVERGENT BOUNDARIES

Divergent boundaries occur where two plates slide apart from each other. Oceanic ridges, which are examples of these divergent boundaries, are where new oceanic, melted lithosphere materials well up, resulting in basaltic magmas that intrude and erupt at the oceanic ridge, in turn creating a new oceanic lithosphere and crust (new ocean floor). Along with volcanic activity, the mid-oceanic ridges are also areas of seismic activity.

TRANSFORM PLATE BOUNDARIES

Transform, or shear/constructive boundaries, do not separate or collide; rather, they slide past each other in a horizontal manner with a shearing motion. Most transform boundaries occur where oceanic ridges are offset on the sea floor. The San Andreas Fault in California is an example of a transform fault.

DID YOU KNOW?

- Plates are in constant motion (several centimeters/yr).
- When collision or grinding occurs, it can create mountains, volcanoes, geysers, and earthquakes.
- Near the junction of these plates are where heat travels rapidly from interior.

ENERGY CONVERSION

The conversion of heat to electricity is common to most power plants. This is the case even though the energy source may be coal, gas, nuclear power, wind power, solar power, waterpower, or geothermal power. Powering the non-transportation section of our economy is important, of course, so the use of any fuel source to convert to electrical power for industrial use is one of our premium objectives in our constant and insatiable appetite for energy. Because of our increasing need for more electricity generation, one might surmise that we ought to focus solely on production of electricity not only to power our economy and industrial complex but also to power everything else—one genie in one bottle to accomplish everything. Liquid fuels are the fuels of choice now because they are accessible, available, and relatively inexpensive. Thus, though our ongoing research for other energy sources persists, we still do not have that absolute pressing need to produce a liquid fuel replacement—not yet we don't.

Anyway, when our energy-needs-focus shifts due to necessity, absolute or otherwise, geothermal is available and we need to continue our research in this important area. Moreover, energy conversion occurs at present with all forms of energy in use and with geothermal conversion is nothing new. Only the procedure and methodology differ. In geothermal energy "conversion" refers to the power plant technology that converts the hot geothermal fluids into electric power.

Even though geothermal power plants have much in common with traditional power-generating stations—turbines, generators, heat exchanges, and other standard power-generating equipment—there are important differences between geothermal and other power-generating technologies. For example, each geothermal site has its own unique set of characteristics and operating conditions. For instance, geothermal sites are essentially site specific; these specific conditions must be considered. That is, the fluid produced from a geothermal well can be steam, brine, or a mixture of the two, and the temperature and pressure of the resource can vary substantially from site to site. The chemical composition of the resource can contain dissolved minerals, gases, and other hard-to-manage substances. Because these site-specific conditions can have a profound effect on efficiency, productivity, and economic viability, engineers strive to fine-tune geothermal conversion technology, precisely matching plant design to the site-specific conditions.

GEOTHERMAL POWER PLANT TECHNOLOGIES

Geothermal power plants fall into one of three conversion categories: direct steam, flash, and binary. The type of conversion used depends on the state of the fluid (whether steam or water) and its temperature. Dry steam power plants were the first type of geothermal power generation plants built. They use the steam from the geothermal reservoir as it comes from wells, and route it directly through turbine/generator units to produce electricity. Flash steam plants are the most common type of geothermal power generation plants in operation today. They use water temperatures greater than 360° F (182° C) that is pumped under high pressure to the generation equipment at the surface. Binary cycle geothermal power generation plants differ from dry steam and flash steam systems in that the water or steam from the geothermal reservoir never meets the turbine/generate units.

DID YOU KNOW?

The temperature of the Earth increases, on average, by about 28° C for every kilometer, or 80° F for every mile, of depth below the surface for the first several kilometers down.

DRY STEAM POWER PLANTS

Location-specific, dry-steam power plants take advantage of subterranean rocks that are so hot that the water from a geothermal power plant vaporizes on its way up through the production well (see Figure 6.1). This type of geothermal power plant can use steam directly from the ground to drive the turbine. This is the oldest type of geothermal power plant. It was first used at Lardarello in Italy in 1904 and is still very effective. These geothermal plants emit only excess steam and a very minor amount of gases (USDOE, 2008).

FLASH STEAM POWER PLANTS

Figure 6.2 is a functional diagram of a flash-steam geothermal power plant. Fluid is sprayed into a tank held at a much lower pressure than the fluid, causing some of the fluid to rapidly vaporize, or "flash." The vapor then drives a turbine, which drives a generator. If any liquid remains in the tank, it is returned to the groundwater pump to be forced down into the Earth again and can be flashed again to extract more energy (USDOE, 2008).

BINARY-CYCLE POWER PLANTS

In a binary-cycle geothermal power plant (see Figure 6.3), water is pumped into the Earth and comes back up hot, just as it does in the flash-steam system. Instead of going into a flash tank, however, the hot water enters a *heat exchanger* (see Figure 6.3), where most of its energy is transferred to another fluid called a *binary liquid*. This fluid can be water, but more often it is a volatile liquid resembling refrigerant that boils easily into vapor at a lower temperature than the water. The liquid-to-vapor conversion occurs in a special low-temperature boiler. Then the vapor leaves the turbine, is cooled back into liquid by a condenser, and is recirculated to the boiler. Because this is a closed-loop system, virtually nothing is emitted into the atmosphere. Moderate temperature water is by far the more common geothermal resource, and most geothermal power plants in the future will be binary-cycle plants (USDOE, 2008).

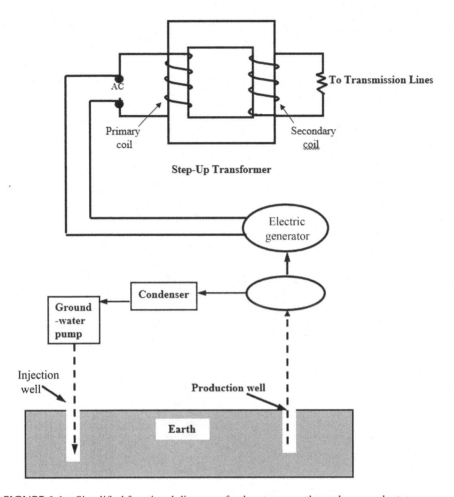

FIGURE 6.1   Simplified functional diagram of a dry steam geothermal power plant.

ENHANCED GEOTHERMAL SYSTEMS (EGS)

The great potential for dramatically expanding the use of geothermal energy can be realized by using enhanced geothermal systems (EGS), also sometimes called engineered geothermal systems. Present geothermal power generation comes from hydrothermal reservoirs and is somewhat limited in geographic application to specific ideal places in the western United States. EERE (2006) points out that this represents the 'lower-hanging fruit' of geothermal energy potential.

EGS offers the chance to extend the use of geothermal resources to large areas of the western United States, as well as into new geographic areas of the entire United States. More than 100,000 Mwe (megawatt electrical) of economically viable capacity may be available in the continental United States, representing a 40-fold increase over present geothermal power generating capacity. This potential is about 10% of the

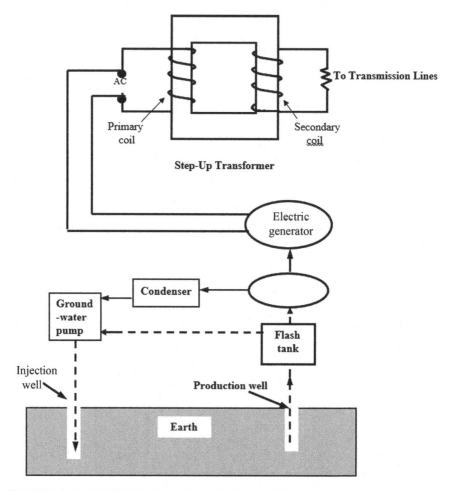

FIGURE 6.2   Simplified functional diagram of a flash-steam geothermal power plant.

overall US electrical capacity today, and as mentioned, represents a domestic energy source that is clean, reliable, and proven.

The EGS concept is to extract heat by creating a subsurface fracture system—a reservoir—to which water can be added through injection wells. A production injection well is drilled into hot basement rock that has limited permeability and fluid content. This type of geothermal resource is sometimes referred to as "hot, dry rock" and represents an enormous potential energy resource. Creating an enhanced, or engineered, geothermal system requires improving the natural permeability of rock. Rocks are permeable due to minute fractures and pore spaces between mineral grains. Injected water is at sufficient pressure to ensure fracturing and is heated by contact with the rock and returns to the surface through production wells, as in naturally occurring hydrothermal systems. Additional production-injection wells are drilled to extract heat from large volumes of rock mass to meet power generation requirements.

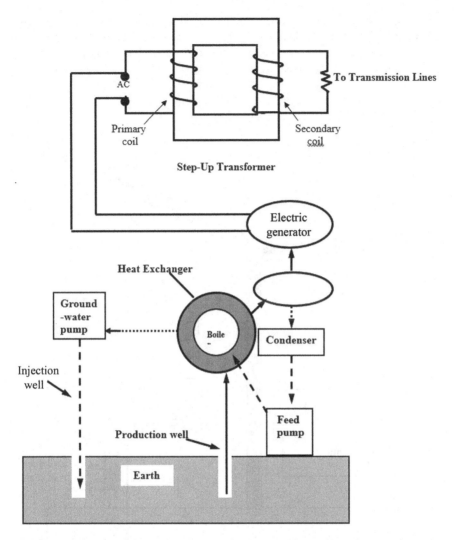

FIGURE 6.3 Simplified functional diagram of a binary-cycle geothermal power plant.

Now a previously unused but large energy resource is available for clean, geothermal power generation.

GEOTHERMAL HEAT PUMPS

Geothermal heat pumps (sometimes referred to as GeoExchange, earth-coupled, ground-source, or water-source heat pumps) have been in use since the late 1940s. Geothermal heat pumps (GHPs) use the constant temperature of the Earth as the exchange medium instead of the outside air temperature. This allows the system to reach high efficiencies (300%–600%) on the coldest of winter nights, compared to 175%–250% for air-source heat pumps on cool days.

Geothermal heat pumps are used for space heating and cooling, as well as water heating. Its great advantage is that it works by concentrating naturally existing heat, rather than by producing heat through combustion of fossil fuels. The system includes three principal components (see Figure 6.4):

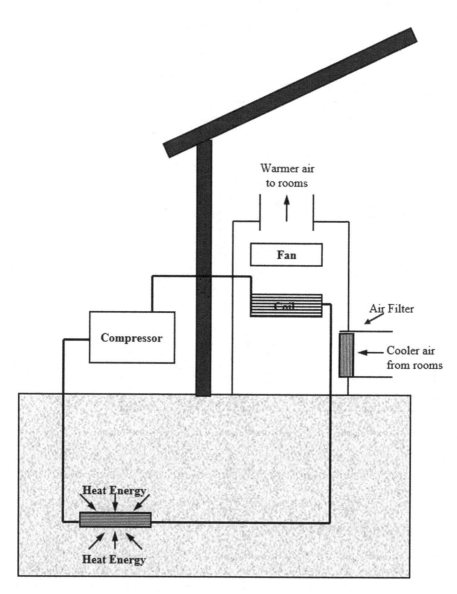

FIGURE 6.4   Geothermal heat pump system.

Source: Adapted from Gibilisco (2007).

- *Geothermal earth connection subsystems*—using the Earth as a heat source/ sink, a series of pipes, commonly called a "loop," is buried in the ground near the building. The loop can be buried either vertically or horizontally. It circulates a fluid (water, or a mixture of water and antifreeze) that absorbs heat from, or relinquishes heat to, the surrounding soil, depending on whether the ambient air is colder or warmer than the soil.
- *Geothermal heat pump subsystem*—for heating, a geothermal heat pump removes the heat from the fluid in the Earth connection, concentrates it, and then transfers it to the building. For cooling, the process is reversed.
- *Geothermal heat distribution subsystem*—conventional ductwork is generally used to distribute heated or cooled air from the geothermal heat pump throughout the building.

In addition to space conditioning, geothermal heat pumps can be used to provide domestic hot water when the system is operating. Many residential systems are now equipped with desuperheaters that transfer excess heat from the geothermal heat pump's compressor to the house's hot water tank. A desuperheater provides no hot water during the spring and fall when the geothermal heat pump system is not operating; however, because the geothermal heat pump is so much more efficient than other means of water heating, manufacturers are beginning to offer "full demand" systems that use a separate heat exchanger to meet all a household's hot water needs. These units cost-effectively provide hot water as quickly any competing system (EERE, 2009).

TYPES OF GEOTHERMAL HEAT PUMPS

Energy Savers (2009) points out that there are four basic types of ground loop geothermal heat pump systems. Three of these—horizontal, vertical, and pond/lade—are closed-loop systems. The fourth type of system is the open-loop option. Which one of these is best depends

DID YOU KNOW?

Geothermal system life is estimated at 25 years for the inside components and 50+ years for the ground loop. There are approximately 50,000 geothermal heat pumps installed in the United States each year. Even with the large number of annual installations, it is important to keep in mind that heat pumps are relatively expensive to install new. This is especially true of the deep ground source type. It may take a long time for a new system to pay for itself.

on the climate, soil conditions, available land, and local installation costs at the site. All these approaches can be used for residential and commercial building applications.

- *Closed-Loop Systems*
  - Horizontal—this type of installation is generally most cost-effective for residential installations, particularly for new construction where sufficient land is available. It requires trenches at least four feet deep. The most common layouts use either two pipes, one buried at six feet, and the other at four feet, or two pipes placed side by side at five feet in the ground in a two-foot-wide trench. The Slinky™ method of looping pipe allows more pipe in a short trench, which cuts down on installation costs and makes horizontal installation possible in areas it would not be with conventional horizontal applications.
  - Vertical—large commercial buildings and schools often use vertical systems because the land area required for horizontal loops would be prohibitive. Vertical loops are also used where the soil is too shallow for trenching, and they minimize the disturbance to existing landscaping. For a vertical system, holes (approximately four inches in diameter) are drilled about 20 feet apart and 100–400 feet deep. Into these holes go two pipes that are connected at the bottom with a U-bend to form a loop. The vertical loops relate to horizontal pipes (i.e., manifold), placed in trenches and connected to the heat pump in the building.
  - Pond/Lake—if the site has an adequate water body, this may be the lowest cost option. A supply line pipe is run underground from the building to the water and coiled into circles at least eight feet under the surface to prevent freezing. The coils should only be placed in a water source that meets minimum volume, depth, and quality criteria.
- Open Loop-System—this type of system uses well or surface body water as the heat exchange fluid that circulates directly through the GHP system. Once it has circulated through the system, the water returns to the ground through the well, a recharge well, or surface discharge. This option is obviously practical only where there is an adequate supply of relatively clean water, and all local codes and regulations regarding groundwater discharge are met.

DID YOU KNOW?

If we utilize waste biomass, solar (passive and thermal), wind (on shore), photovoltaic, geothermal and other renewable resources available to us in the United States, we will exceed the demand (what we need) by at least five times as much energy as we need, all from clean, renewable sources.

–Frank R. Spellman, 2010, p. 12

THE BOTTOM LINE

The supply of geothermal energy is vast and can be considered renewable if each site is properly engineered and operated to ensure that excessive water is not pumped into the Earth in one location in too short a time. Geothermal energy can be and already is accessed by drilling water or steam wells in a process that is like drilling for oil. Geothermal energy is an enormous, underused heat and power resource that is clean (emits little or no greenhouse gases), reliable (average system availability of 95%), and homegrown (making us less dependent on foreign oil).

Geothermal resources range from shallow ground to hot water and rocks several miles below the Earth's surface, and even farther down to the extremely hot molten rock called magma. Mile-or-more-deep wells can be drilled into underground reservoirs to tap steam and very hot water that can be brought to the surface for use in a variety of applications. In the United States, most geothermal reservoirs are in the western states, Alaska, and Hawaii. However, before geothermal electricity can be considered a key element of the US energy infrastructure, it must become cost-competitive with traditional forms of energy.

According to US Environmental Protection Agency (2022), geothermal heat pumps are the most energy-efficient, environmentally clean, and coal-effective systems for heating and cooling buildings. All types of buildings, including homes, office buildings, schools, and hospitals, can use geothermal heat pumps.

NOTE

1   Based on information from USDOE (2001). *Renewable Energy: An Overview*. Washington, DC: US Department of Energy.

REFERENCES AND RECOMMENDED READING

Energy Efficiency & Renewable Energy (EERE). 2006. *How an Enhanced Geothermal System Works*. Accessed 03/29/10 @ www1.eere.energy.gov/gelotherm/printable_versions/egs_animation.html

Energy Efficiency & Renewable Energy (EERE). 2009. *Geothermal Heat Pumps*. Accessed 03/27/10 @ www1.eere.energy.gov/ gelothermal/heatpumps.html?print

Energy Savers. 2009. *Types of Geothermal Heat Pump Systems*. Washington, DC: US Department of Energy. Accessed @ www.energysavers.gov/your_home/space _heating _ cooling/index

Gibilisco, S. 2007. *Alternative Energy Demystified*. New York: McGraw Hill.

Kutsher, C.F. 2000. *The Status and Future of Geothermal Electric Power*. Accessed 4/4/23 @ https://api.semanticscholor.org/copiesID:15012696

Lund, J.W. 2007. Characteristics, development and utilization of geothermal sources. *Geo-Heat Centre Quarterly Bulletin* **28**(2): 1–9.

McLarty, L., Grabowski, P., Entingh, D. and Robertson-Tait, A. 2000. Enhanced geothermal systems R&D in the United States. WGC 2000. pp. 3793–96.

Spellman, F.R. 2010. *The Science of Renewable Energy*. Boca Raton, FL: CRC Press.

US Department of Energy (USDOE). 2006. *Geothermal Technologies Program: A History of Geothermal Energy in the United States*. Accessed 06/18/09 @ www1.eere.energy.gov/geothermal/printableversison /history.html

US Department of Energy (USDOE). 2008. *Hydrothermal Power Systems*. Accessed 03/26/10 @ www1.eere.energy.gov. geothermal/printable_versions/powerplants.html

US Department of Energy (USDOE). 2009. *Fossil Fuels*. US Dept of Energy. Accessed 06/10/
     09 @ www.energy.gov/ energysources/fossilfuels.htm

US Environmental Protection Agency (USEPA). 2022. *Geothermal Explained: Geothermal
     Heat Pumps*. Accessed 6/29/2 @ www.eia.gov/energyexplained/geothermal-heat-
     puts.php

US Geological Survey (USGS). 2008. *Assessment of Moderate- and High-Temperature
     Geothermal Resources of the United States*. Accessed 6/28/23 @ http://pubs.usgs.gov/
     fs/2008

7 Marine Energy

THE RIPPLE EFFECT

Also known as marine and hydrokinetic energy or marine renewable energy, *marine energy* is a green power source that is harnessed from the natural movement of water, including waves, tides, ripples, and river and ocean currents. Marine energy can be harnessed from temperature differences in water through a process known as ocean thermal energy conversion.

SIDEBAR 7.1—THE SEA PAYS NO HOMAGE TO KINGS

One day King Canute told his followers to carry him and his throne to the seashore at low tide. "Set me down right there at the water's edge," he said. "And I will command the sea to stay away. I will order the tide not to rise."

"Oh wow!" said his followers as they obeyed the King. "This ought to be something to see!"

King Canute sat. He shouted: "Sea, stay away! Tide, do not rise!" But slowly, slowly, the tide came in and the sea rose, over his feet, past his knees, up to his waist. The crowd of followers pulled back a little to keep their own feet dry. They were puzzled.

Why did the sea not obey the King?

The water rose higher and higher, up to the King's shoulders, over his chin. "Loyal glub subjects!" he gurgled. "Now you see that glub something no man, be he glub king or commoner, can glub do! Now pull me the glub out of here!"

–Based on the classic fairy tale

As King Canute reportedly discovered, the rise and fall of the seas represents a vast and relentless natural phenomenon—certainly beyond the absolute control of all earthly subjects.

The ocean can produce two types of energy: *thermal energy* from the sun's heat, and *mechanical energy* from the tides and waves. Generating technologies for

DOI: 10.1201/9781003439059-8

deriving electrical power from the ocean include tidal power, wave power, ocean thermal energy conversion, oceans currents, ocean winds, and salinity gradients. Of these, the three most well-developed technologies are tidal power, wave power, and ocean thermal energy conversion (OTEC). Tidal power requires large tidal differences which, in the United States, occur only in Maine and Alaska. Ocean thermal energy conversion is limited to tropical regions, such as Hawaii, and to a portion of the Atlantic coast. Wave energy has a more general application, with potential along the California coast. The western coastline has the highest wave potential in the United States, and in California, the greatest potential is along the northern coast. Ocean thermal energy can be used for many applications, including electricity generation. Electricity conversion systems use either the warm surface water or boil seawater to turn a turbine, which activates a generator.

The electricity conversion of both tidal and wave energy usually involves mechanical devices. It is important to distinguish tidal energy from hydropower. Recall that hydropower is derived from the hydrological climate cycle, powered by solar energy, which is usually harnessed via hydroelectric dams. In contrast, tidal energy is the result of the interaction of the gravitational pull of the moon and, to a lesser extent, the sun, on the seas. Processes that use tidal energy rely on the twice-daily tides, and the resultant upstream flows and downstream ebbs in estuaries and the lower reaches of some rivers, as well, in some cases, tidal movement out at sea. A dam is typically used to convert tidal energy into electricity by forcing the water through turbines, activating a generator. Meanwhile, wave energy, a very large potential resource to be tapped, uses mechanical power to directly activate a generator, to transfer to a working fluid, water, or air, which then drives a turbine/generator.

KEY TERMS AND DEFINITIONS[1]

Waves

- *Point Absorber*—is a floating or submerged energy capture device, with principal dimension relatively small compared to the wavelength, and can capture energy from a wave front greater than the physical dimension of the device.
- *Submerged Pressure Differential*—a fully submerged point absorber used to capture wave energy; a pressure differential is induced with the device as the wave passes, driving a fluid pump to create mechanical energy.
- *Oscillating Water Column*—these shore-based or floating partially submerged structures enclose a column of air above a column of water. A collector funnels waves into the structure below the waterline, causing the water column to rise and fall; this alternately pressurizes and depressurizes the air column, pushing or pulling it through a turbine.
- *Overtopping Device*—these shore- and floating models are partially submerged structures. A collector funnels waves over the top of the structure into a reservoir; water runs back out to the sea from this reservoir thought a turbine.

- *Attenuator*—this is a wave capture device with principal axis oriented parallel to the direction of the incoming wave that converts the energy due to relative motion of the parts of the device as the wave passes along it.
- *Oscillating Wave Surge Converter*—these devices capture wave energy directly without a collector by using relative motion between a float/flap/membrane and a fixed reaction point. The float/flap/membrane oscillates along a given axis dependent on the device; mechanical energy is extracted from the relative motion of the body part relative to its fixed reference.

Current

- *Axial Flow Turbine*—these shrouded or open rotor turbines are oriented in the direction of water flow where the kinetic motion of the water current creates lift on blades causing the rotor to turn, driving a mechanical generator.
- *Cross Flow Turbine*—these devices typically have two or three blades mounted along a vertical shaft to form a rotor; the kinetic motion of the water current creates lift on the blades causing the rotor to turn, driving a mechanical generator. These turbines can operate with flow from multiple directions without reorientation.
- *Reciprocating Device*—produces a vortex, the Magnus effect, or by flow flutter while using the flow of water to produce lift or drag of an oscillating part transverse to the flow direction. The oscillating hydrofoil is like an airplane but in water. Yaw control systems adjust their angle relative to the water stream, creating lift and drag forces that cause device oscillation. Mechanical energy from this oscillation feeds into a power conversion system.

Ocean Thermal Energy Conversion (OTEC)

- *Closed-cycle*—these systems use fluid with a low boiling point, such as ammonia, to rotate a turbine to generate electricity. Warm surface seawater is pumped through a heat exchanger where the low boiling point vaporized. The expanding vapor turns the turbo generator. Cold deep-seawater—pumped through a second heat exchanger—condenses the vapor back into a liquid, which is then recycled through the system.
- *Open-cycle*—these systems use the tropical oceans' warm surface water to make electricity. When warm seawater is placed in a low-pressure container, it boils. The expanding steam drives a low-pressure electrical turbo-generator. The steam, which has left its salt behind in the low-pressure container, is almost pure fresh water. It is condensed back into a liquid by exposure to cold temperatures from deep-ocean water.
- *Hybrid*—these systems combine the features of both the closed-cycle and the open-cycle system, warm seawater system, warm seawater enters a vacuum chamber where it is flash-evaporated into steam, like the open-cycle evaporation process. The steam vaporizes a low-boiling-point fluid (in a closed-cycle loop) that drives a turbine to produce electricity.

Note: Before beginning a detailed discussion of the ocean's thermal and mechanical energy potential, it is important to have a basic understanding of oceans, and especially their margins; it is at the margins of coastal regions where most, if not all, ocean energy is harnessed using present technology. Accordingly, the following section is provided as foundational information for better understanding the ocean green energy concepts presented in this chapter.

OCEANS AND THEIR MARGINS[2]

Oceans are a principal component of the hydrosphere and the storehouse of Earth's water. Oceans cover about 71% of Earth's surface. The average depth of Earth's oceans is about 3,800 m, with the greatest ocean depth recorded at 11,036 m in the Mariana Trench. At the present time, the oceans contain a volume of about 1.35 billion cubic kilometers (96.5 % of Earth's total water supply), but the volume fluctuates with the growth and the melting of glacial ice.

Composition of ocean water has remained constant throughout geological time. The major constituents dissolving in ocean water (from rivers and precipitation, and the result of weathering and degassing of the mantle by volcanic activity) is composed of about 3.5%, by weight, of dissolved salts including chloride (55.07%), sodium (30.62%), sulfate (7.72%), magnesium (3.68%), calcium (1.17%), potassium (1.10%), bicarbonate (0.40%), bromine, (0.19%), and strontium (0.02%).

The most significant factor related to ocean water that everyone is familiar with is the salinity of the water—how salty it is. *Salinity*, a measure of the amount of dissolved ions in the oceans, ranges between 33 and 37 parts per thousand. Often the concentration is the amount (by weight) of salt in water, as expressed in "parts per million" (ppm—analogous to a full whiskey shot glass of water taken from an Olympic-sized swimming pool). Water is saline if it has a concentration of more than 1,000 ppm of dissolved salts; ocean water contains about 35,000 ppm of salt (USGS, 2007). Chemical precipitation, absorption into clay minerals, and plants and animals prevent seawater from containing even higher salinity concentrations. However, salinity does vary in the oceans because surface water evaporates, rain and stream water is added, and ice forms or thaws.

Along with salinity, another important property of seawater includes temperature. The temperature of surface seawater varies with latitude, from near 0° C near the poles to 29° C near the equator. Some isolated areas can have temperatures up to 37° C. Temperature decreases with ocean depth.

THE OCEAN FLOOR

The bottoms of the oceans' basins (ocean floors) are marked by mountain ranges, plateaus, and other relief features like (although not as rugged as) those on the land.

As shown in Figure 7.1, the floor of the ocean has been divided into four divisions: the continental shelf, the continental slope, the continental rise, and the deep-sea floor or abyssal plain.

- *Continental Shelf*—this is the flooded, nearly flat, true margins of the continents. Varying in width to about 40 miles and a depth of approximately 650 feet, continental shelves slope gently outward from the shores of the continents (see Figure 7.1). Continental shelves occupy approximately 7.5% of the ocean floor.
- *Continental Slope*—this is a relatively steep slope descending from the continental shelf (see Figure 7.1) that descends rather abruptly to the deeper parts of the ocean. These slopes occupy about 8.5% of the ocean floor.
- *Continental Rise*—this is a broad gentle slope below the continental slope containing sediment that has accumulated along parts of the continental slope.
- *Abyssal Plain*—this is a sediment-covered deep-sea plain about 12,000–18,000 feet below sea level. This plane makes up about 42% of the ocean floor.

The deep ocean floor does not consist exclusively of the abyssal plain. In places, there are areas of considerable relief. Among the more important such features are:

- *Seamounts*—these are isolated mountain-shaped elevations more than 3,000 feet high.
- *Mid-oceanic Ridge*—these are submarine mountains, extending more than 37,000 miles through the oceans, generally 10,000 feet above the abyssal plain.
- *Trench*—this is a deep, steep-sided trough in an abyssal plain.
- *Guyots*—this is a seamount that is flat-topped and was once a volcano. They rise from the ocean bottom and usually are covered by 3,000 to 6,000 feet of water.

OCEAN TIDES, CURRENTS, AND WAVES

Water is the expert sculptor of Earth's surfaces. The ceaseless, restless motion of the sea is an extremely effective geological agent. Besides shaping inland surfaces,

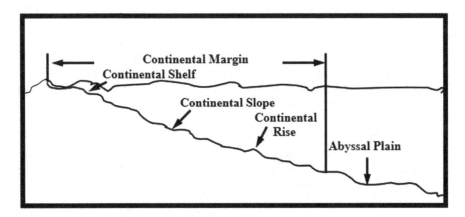

FIGURE 7.1   Cross section of ocean floor showing major elements of topography.

water sculpts the coast. Coasts include sea cliffs, shores, and beaches. Seawater set in motion erodes cliffs, transport erodes debris along shores, and dumps it on beaches. Therefore most coasts retreat or advance. In addition to the unceasing causes of motion—wind, density of sea water, and rotation of the Earth—the chief agents in this process are tides, currents, and waves.

Tides

The periodic rise and fall of the sea (once every 12 hours and 26 minutes) produces the tides. Tides are due to the gravitational attraction of the moon and to a lesser extent, the sun on the Earth. The moon has a larger effect on tides and causes the Earth to bulge toward the moon. It is interesting to note that at the same time the moon causes a bulge on Earth, a bulge occurs on the opposite side of the Earth due to inertial forces (further explanation is beyond the scope of this text). The effect of the tides is not too noticeable in the open sea, the difference between high and low tide amounting to about two feet. The tidal range may be considerably greater near shore, however. It may range from less than 2 feet to as much as 50 feet. The tidal range will vary according to the phase of the moon and the distance of the moon from the Earth. The type of shoreline and the physical configuration of the ocean floor will also affect the tidal range.

Currents

The oceans have localized movements of masses of seawater called ocean currents. These are the result of drift of the upper 50 to 100 m of the ocean due to drag by wind. Thus, surface ocean currents generally follow the same patterns as atmospheric circulation, with the exception that atmospheric currents continue over the land surface, while ocean currents are deflected by the land. Along with wind action, current may also be caused by tides, variation in the salinity of the water, rotation of the Earth, and concentrations of turbid or muddy water. Temperature changes in water affect water density, which, in turn, causes currents—these currents cause seawater to circulate vertically.

Waves

Waves, varying greatly in size, are produced by the friction of wind on open water. Wave height and power depend upon wind strength and fetch—the amount of unobstructed ocean over which the wind has blown. In a wave, water travels in loops. Essentially an up-and-down movement of the water, the diameter of the loops decreases with depth. The diameter of loops at the surface is equal to wave height (h). Breakers are formed when the wave comes into shallow water near the shore. The lower part of the wave is retarded by the ocean bottom, and the top, having greater momentum, is hurled forward, causing the wave to break. These breaking waves may do great damage to coastal property as they race across coastal lowlands driven by wind or gale or hurricane velocities.

COASTAL EROSION, TRANSPORTATION, AND DEPOSITION

The geological work of the sea, like previously discussed geological agents, consists of erosion, transportation, and deposition. The sea accomplishes its work of coastal landform sculpting largely by means of waves and wave-produced currents; their effect on the seacoast may be quite pronounced. The coast and accompanying coastal deposits and landform development represent a balance between wave energy and sediment supply.

Wave Erosion
Waves attack shorelines and erode by a combination of several processes. The resistance of the rocks composing the shoreline and the intensity of wave action to which it is subjected, are the factors that determine how rapidly the shore will be eroded. Wave erosion works chiefly by hydraulic action, corrosion, and attrition. As waves strike a sea cliff, *hydraulic action* crams air into rock crevices, putting tremendous pressure on the surrounding rock; as waves retreat, the explosively expanding air enlarges cracks and breaks off chunks of rock (*scree*). Chunks hurled by waves against the cliff break off more scree (via a sandpapering action)—a process called *corrasion*. When the sea rubs and grinds rocks together forming scree, it is thrown into the cliffs, reducing broken rocks to pebbles and sand grains—a process called *attrition* (Lambert, 2007).

Several features are formed by marine erosion—different combinations of wave action, rock type, and rock beds produce these features. Some of the more typical erosion-formed features of shorelines are discussed below.

- *Sea Cliffs or Wave-Cut Cliffs*—these are formed by wave erosion of underlying rock followed by the caving-in of the overhanging rocks. As waves eat farther back inland, they leave a wave-cut beach or platform. Such cliffs are essentially vertical and are common at certain localities along the New England and Pacific coasts of North America.
- *Wave-cut Bench*—these are the result of wave action not having enough time to lower the coastline to sea level. Because of the resistance to erosion, a relatively flat wave-cut bench develops. If subsequent uplift of the wave-cut bench occurs, it may be preserved above sea level as a wave-cut bench.
- *Headlands*—these are finger-like projections of resistant rock extending out into the water. Indentations between headlands are termed *coves*.
- *Sea Caves, Sea Arches, and Stacks*—these are formed by continued wave action on a sea cliff. Wave action hollows out cavities or caves in the sea cliffs. Eventually, waves may cut completely through a headland to form a sea arch; if the roof of the arch collapses, the rock that is left separated from the headland is called a stack.

MARINE TRANSPORTATION

Waves and currents are important transporting agents. Rip currents and undertow carry rock particles back to the sea, and long-shore currents will pick up sediments (some of it in solution), moving them out from shore into deeper water. Materials carried in solution or suspension may drift seaward for great distances and eventually be deposited far from shore. During the transportation process, sediments undergo additional erosion, becoming reduced in size.

MARINE DEPOSITION

Marine deposition takes place whenever currents and waves suffer reduced velocity. Some rocks are thrown up on the shore by wave action. Most of the sediments thus deposited consist of rock fragments derived from the mechanical weathering of the continents, and they differ considerably from terrestrial or continental deposits. Due to input of sediments from rivers, deltas may form, and as a result of beach drift, such features as spits and hooks, bay barriers, and tombolos may form. Depositional features along coasts are discussed below.

- *Beaches*—these are transitory coastal deposits of debris that lie above the low-tide limit in the shore zone.
- *Barrier Islands*—these are long narrow accumulations of sand lying parallel to the shore and separated from the shore by a shallow lagoon.
- *Spits and Hooks*—these are elongated, narrow embankments of sand and pebble extending out into the water but attached by one end to the land.
- *Tombolos*—these are bars of sand or gravel connecting an island with the mainland or another island.
- *Wave-built Terraces*—these are structures built up from sediments deposited in deep water beyond a wave-cut terrace.

WAVE ENERGY[3]

Waves are caused by the wind blowing over the surface of the ocean. In many areas of the world, the wind blows with enough consistency and force to provide continuous waves. Wave energy does not have the tremendous power of tidal fluctuations, but the regular pounding of the waves should not be underestimated because there is tremendous energy in the ocean waves. The total power of waves breaking on the world's coastlines is estimated at between 2 and 3 million megawatts. In optimal wave areas, more than 65 megawatts of electricity could be produced along a single mile of shoreline, according to EERE (2004). In essence, because the wind is originally derived from the sun, we can consider the energy in ocean waves to be a stored, moderately high-density form of solar energy. According to certain estimates, wave technologies could feasibly fulfill 10% of the global electricity supply if fully developed (World Energy 2004). The west coasts of the United States and Europe and the coast of Japan and New Zealand are good sites for harnessing wave energy.

WAVE ENERGY: FACTS, PARAMETERS, AND EQUATIONS

- Three main processes create waves: air flowing over the sea exerts a tangential stress on the water surface, resulting in the formation and growth of waves; turbulent air flow close to the water surface creates rapidly varying shear stresses

and pressure fluctuations (when these oscillations are in phase with existing waves, further wave development occurs); and when waves have reached a certain size, the wind can exert a stronger force on the up-wind face of the wave, resulting in additional wave growth.
- Waves located within or close to the areas where they are generated, are called storm waves.
- *Swell waves* can develop at great distances from the point of origin.
- The distance over which wind energy is transferred into the ocean to form waves is called the *fetch*.
- *Sea state* is the general condition of the free surface on a large body of water—with respect to wind waves and swell—at a certain location and moment. The World Meteorological Organization sea state code largely adopts the "wind sea" definition of the Douglas Sea Scale.

WMO Sea State Code	Wave Height (meters)	Characteristics
0	0	Calm (glassy)
1	0 to 0.1	Calm (rippled)
2	0.1 to 0.5	Smooth (wavelets)
3	0.5 to 1.25	Slight
4	1.25 to 2.5	Moderate
5	2.5 to 4	Rough
6	4 to 6	Very Rough
7	6 to 9	High
8	9 to 14	Very high
9	Over 14	Phenomenal

- The shape of a typical wave is described as *sinusoidal* (that is, it has the form of a mathematical sine function; see Figure 7.2). The difference in height between the peaks troughs is known as the *height*, H, and the distance between successive peaks (or troughs) of the wave is known as the wavelength, λ. The time in seconds taken for successive peaks (or troughs) to pass a given fixed point is known as the *period*, T. The *frequency*, v, of the wave describes the number of peak-to-peak (or trough-to-trough) oscillations of the wave surface per second, as seen by a fixed observer, and is the reciprocal of the period. That is, $v = 1/T$.

If a wave is traveling at velocity v past a given fixed point, it will travel a distance equal to its wavelength λ in a time equal to the wave period T (i.e., $v = \lambda/T$). The power, P, (Kw/m) of an idealized ocean wave is approximately equal to the square of the height, H (meters), multiplied by the wave period, T (seconds). The exact expression is the following:

$$P = \frac{g^2 H^2 T}{32},$$

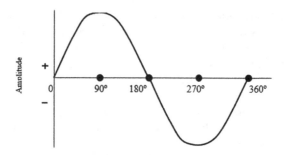

FIGURE 7.2 Sinusoidal wave showing wavelength and amplitude.

where P is in units of watts/m and g is the acceleration due to gravity (9.81 m s^{-2}) (Phillips, 1977).

- Deep water waves—the velocity of a long ocean wave—can be shown to be proportional to the period (if the depth of water is greater than about half of the wavelength λ) as follows:

$$v = \frac{gT}{2\pi}$$

A useful approximation can be derived from this: velocity in meters/second is about 1.5 times the wave period in seconds. The result leads to deep ocean waves traveling faster than the shorter waves. Moreover, if the above relationships hold, we can find the deep-water wavelength, λ, for any given wave period.

$$\lambda = \frac{gT}{2\pi}$$

- Intermediate depth waves—as the water becomes shallower, the properties of the waves become increasingly dominated by water depth. When waves reach shallow water, their properties are completely governed by the water depth, but in intermediate depths (i.e., between $d = \lambda/4\pi$), the properties of the waves will be influenced by both water depth d and wave period T (Phillips, 1977; Goda, 2000).
- Shallow water waves—as waves approach the shore, the seabed starts to influence their speed, and it can be shown that if the water depth d is less than a quarter of the wavelength, the velocity is given by:

$$V = \sqrt{gd}.$$

- As waves propagate, their energy is transported. The energy transport velocity is the group velocity. As a result, the wave energy flux, through a vertical plane of unit width perpendicular to the wave propagation direction, is equal to:

$$P = E\,c_g,$$

where c_g is the group velocity (m/s).

DID YOU KNOW?

The wave height increases with increases in:

- wind speed
- time duration of the wind blowing
- fetch
- water depth

WAVE ENERGY CONVERSION TECHNOLOGY

In the early 1970s, the harnessing of wave power focused on using floating devices such as Cockerell Rafts (a wave power hydraulic device), the Salter Duck (curved cam-like device that can capture 90% of waves for energy conversion), the Rectifier (concerts A-C to D-C electricity), and the Clam (a floating rigid toroid—i.e., doughnut-shaped—that converts wave energy to electrical energy). Wave energy converters can be classified in terms of their location: fixed to the seabed, generally in shallow water; floating offshore in deep water; or tethered in intermediate depths. At present, these floating devices are not cost effective and have very difficult moving problems. So current practice is to move inshore, sacrificing some energy, but fixed devices, according to Tovey (2005), have several advantages, including:

- easier maintenance
- easier to land on device
- no mooting problem
- easier power transmission
- enhanced productivity
- better design life

DID YOU KNOW?

Hydro (water) Kinetic Energy is the energy possessed by a body of water because of its motion ($KE = 1.2\ mv^2$). Hydro-Static (at rest) Energy is the energy possessed by a body because of its position or location at an elevation or height above a reference or datum ($PE = mgh$)

Wave energy devices can be classified by means of their reaction system, but it is often more instructive to discuss how they interact with the wave field. In this context, each moving body may be listed as either displace or reactor.

- Displacer—this is the body moved by the waves. It might be a buoyant vessel or a mass of water. If buoyant, the displacer may pierce the surface of the waves or be submerged.
- Reactor—this is the body that provides reaction to the displacer. As suggested above, it could be a body fixed to the seabed, or the seabed itself. It could also be another structure or mass that is not fixed but moves in such a way that reaction forces are created (e.g., by moving by a different amount or at different times). A degree of control over the forces acting on each body and/or acting between the bodies (particularly stiffness and damping characteristics) is often required to optimize the amount of energy captured.

Note: In some designs, the reactor is inside the displacer, while in others it is an external body. Internal reactors are not subject to wave forces, but external ones may experience loads that cause them to move in ways like a displacer. This can be extended to the view that some devices do not have dedicated reactors at all, but rather a system of displacers whose relative emotion creates a reaction system (INEL, 2005).

There are three types of well-known Wave Energy Conversion devices: point absorbers, terminators, and attenuators (see Figure 7.3; INEL, 2005).

- Point absorber—this is a floating structure that absorbs energy in all directions by virtue of its movements at or near the water surface (see Figure 7.4). It

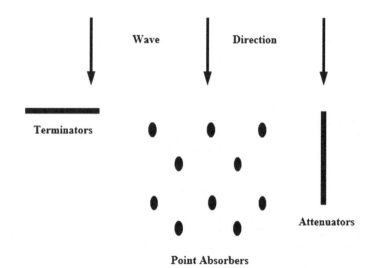

FIGURE 7.3   Types of wave energy converter.

Source: Adaptation from N.K. Tovey, 2005.

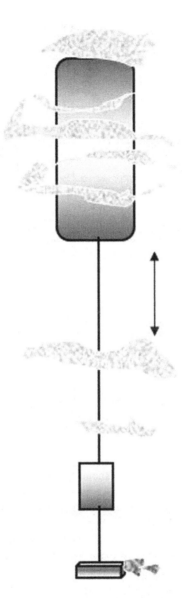

FIGURE 7.4 Point absorber.

may be designed to resonate—that is, move with larger amplitudes than the waves themselves. This feature is useful to maximize the amount of power that is available for capture. The power take-off system may take several forms, depending on the figuration of displacers/reactors.

- Terminator—this is also a floating structure that moves at or near the water surface, but it absorbs energy in only a single direction (see Figure 7.5). The device extends in the direction normal to the predominant wave direction, so that as

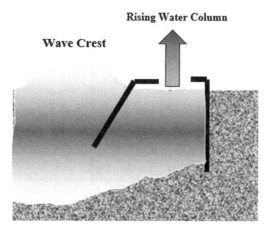

FIGURE 7.5   Terminator.

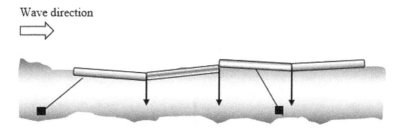

FIGURE 7.6   Attenuator.

waves arrive, the device restrains them. Again, resonance may be employed and the power take-off system may take a variety of forms.

- Attenuator—this device is a long floating structure like the terminator but is oriented parallel to the waves rather than normal to them (see Figure 7.6). It rides the waves like a ship, and movements of the device at its bow and along its length can be restrained to extract energy. A theoretical advantage of the attenuator over the terminator is that its area normal to the waves is small and therefore the forces it experiences are much lower.

TIDAL ENERGY

The tides rise and fall in eternal cycles. Tides are changes in the level of the oceans caused by the gravitational pull of the moon and sun, and the rotation of the Earth. The relative motions of these cause several different tidal cycles, including: a semi-diurnal cycle—period 12 hrs 25 minutes; a semi-monthly cycle—that is, Spring or Neap Tides corresponding with the position of the moon; a semi-annual cycle—period of about 178 days, which is associated with the inclination of the Moon's orbit. This causes the highest spring tides to occur in March and September; and other long-term cycles—for example, a 19-year cycle of the moon. Nearshore water levels can vary up to 40 feet, depending on the season and local factors. Only about 20 locations have good inlets and a large enough tidal range—about 10 feet—to produce energy economically (USDOE, 2010).

DID YOU KNOW?

The spring tides have a range about twice that of neap tides, while the other cycles can cause further variations of up to 15%. The tidal range is amplified in estuaries, and in some situations, the shape of the estuary is such that near resonance occurs.

TIDAL ENERGY TECHNOLOGIES

Some of the oldest ocean energy technologies use tidal power. Tidal power is more predictable than solar power and wind energy. All coastal areas consistently experience two high and two low tides over a period slightly greater than 24 hours. For those tidal differences to be harnessed into electricity, the difference between high and low tide must be at least five meters, or more than 16 feet. There are only about 40 sites on the Earth with tidal ranges of this magnitude. Currently, there are no tidal power plants in the United States. However, conditions are good for tidal power generation in both the Pacific Northwest and the Atlantic Northeast regions of the country. Tidal energy technologies include the following:

- Tidal Barrages—a barrage or dam is a simple generation system for tidal plants that involves a dam, known as a barrage, across an inlet. Sluice gates (gates commonly used to control water levels and flow rates) on the barrage allow the tidal basin to fill on the incoming high tides and to empty through the turbine system on the outgoing tide, also known as the ebb tide. There are two-way systems that generate electricity on both the incoming and outgoing tides. A potential disadvantage of a barrage tidal power system is the effect a tidal station can have on plants and animals in estuaries. Tidal barrages can change the tidal level in the basin and increase the amount of matter in suspension in the water (turbidity). They can also affect navigation and recreation.
- Tidal fences—these look like giant turnstiles. A tidal fence has vertical axis turbines mounted in a fence. All the water that passes is forced through the turbines. Some of these currents run at 5–8 knots (5.6–9 miles per hour) and generate as much energy as winds of much higher velocity. Tidal fences can be used in areas such as channels between two landmasses. Tidal fences are cheaper to install than tidal barrages and have less impact on the environment tidal barrages, although they can disrupt the movement of large marine animals.
- Tidal turbine—are basically wind turbines in the water that can be located anywhere there is a strong tidal flow (they function best where coastal

currents run at between 3.6 and 4.9 knots—4 to 5.5 mph). Because water is about 800 times denser than air, tidal turbines must be much sturdier than wind turbines. Tidal turbines are heaver ad more expensive to build but capture more energy.

OCEAN THERMAL ENERGY CONVERSION

As mentioned, the most plentiful renewable energy source on our planet by far is solar radiation: 170,000 TW (170,000 x 10^{12} W) fall on Earth. Because of its dilute and erratic nature, however, it is difficult to harness. To do so, that is, to capture this energy, we must employ the use of large collecting areas and large storage capacities; these requirements are satisfied on Earth only by the tropical oceans. We are all taught at an early age that oceans (and water in general) cover about 71% (or two-thirds) of Earth's surface. In a fitting reference to the vast oceans covering most of the Earth, Ambrose Bierce (1842–1914) commented: "A body of water occupying about two-thirds of the world made for man who has no gills." So, true, we have no gills; thus, for those who look out upon those vast bodies of water that cover the surface, they might ask: What is their purpose? And, of course, this is a good question with several possible answers. Regarding renewable energy, we can look out upon those vast seas and wonder: How can we use this massive storehouse of energy for our own needs? Because it is so vast and deep, it absorbs much of the heat and light that come from the sun. One thing seems certain: Our origin, past, present, and future, lies within those massive wet confines we call oceans.

OCEAN ENERGY CONVERSION PROCESS[4]

The ocean is essentially a gigantic solar collector. The energy from the sun heats the surface water of the ocean. In tropical regions, the surface water can be 40 or more degrees warmer than the deep water. This temperature difference can be used to produce electricity. Ocean Thermal Energy Conversion (OTEC) has the potential to produce more energy than tidal, wave, and world energy combined.

The OTEC systems can be open or closed. In a closed system, an evaporator turns warm surface water into stream under pressure (see Figures 7.7 and 7.8). This steam spins a turbine generator to produce electricity. Water pumps bring cold deep water though pipes to a condenser on the surface. The cold water condenses the steam, and the closed cycle begins again. In an open system, the steam is turned into fresh water, and new surface water is added to the system. A transmission cable carries the electricity to the shore.

The OTEC systems must have a temperature difference of about 25° C to operate. This limits OTEC's use to tropical regions where the surface waters are very warm and there is deep cold water. Hawaii, with its tropical climate, has experimented with OTEC systems since the 1970s.

Because there are many challenges to widespread use, there are no large or major operations, but, at present, there are several experimental OTEC plants. Pumping the water is a giant engineering challenge. Because of this, OTEC systems are not very

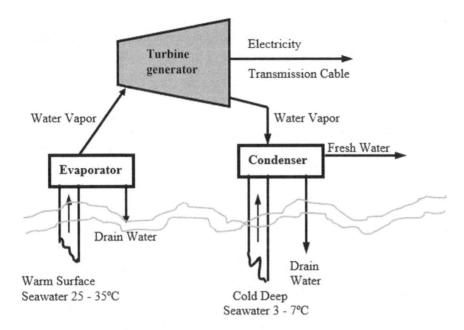

FIGURE 7.7 Schematic of ocean thermal energy conversion system.

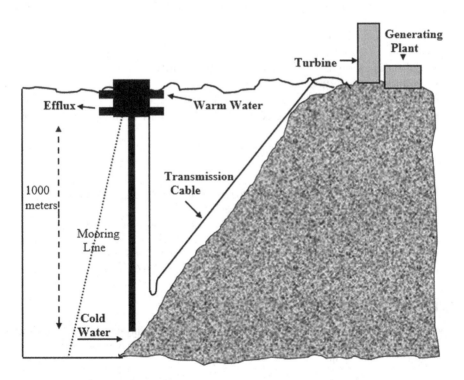

FIGURE 7.8 OTEC floating platform.

energy efficient. The USDOE (2010) estimates it will probably be 10 to 20 years before the technology is available to produce and transmit electricity economically from OTEC systems.

THE BOTTOM LINE

The three forms of ocean energy—tidal, wave, and OTEC systems—are all renewable, clean sources of energy. This is a significant advantage over fossil fuels and other energy forms that pollute the environment. Although all three forms of ocean energy show promise for future development, it is the OTEC systems that appear to be most beneficial for use at the present time. OTEC systems provide both economic and noneconomic benefits.

Economic benefits include:

- Helps produce fuels such as hydrogen, ammonia, and methanol
- Provides moderate-temperature refrigeration
- Produces baseload electrical energy
- Produces desalinated water for industrial, agricultural, and residential uses
- Is a resource for onshore and near-shore mariculture operations
- Provides air conditioning to buildings
- Has significant potential to provide clean, cost-effective electricity for the future

Noneconomic benefits include:

- Enhances energy independence and energy security
- Promotes competitiveness and international trade
- Promotes international sociopolitical stability
- Has potential to mitigate greenhouse gas emissions resulting from burning fossil fuels

NOTES

1 Based on definitions provided in EERE (2022) *Advantages of Marine Energy*. Washington, DC: Energy and Efficiency & Renewable Energy.
2 Material in this section is adapted from F.R. Spellman's *Geology for Non-Geologists*. Lanham, MD: Government Institutes Press.
3 From DOI (2010) *Ocean Energy*. Accessed 03/31/10 @ www.mms.gov.
4 Much of the information in this section is from USDOE (2010) *Ocean Energy*.

REFERENCES AND RECOMMENDED READING

Goda, Y. 2000. *Random Seas and Designs of Maritime Structures*. Singapore: World Scientific. Accessed 04/01/23 @ www.esru.strath.ac.uk/EandE/Web_sites/01-02/RE_info/wave%20power.htm

Idaho National Engineering Laboratory (INEL). 2005. *DOE HydroKinetic Workshop*. Accessed 04/06/22 @ www.hydropower.inel.gov/hydrokinetic_wave.pdf

Lambert, D. 2007. *The Field Guide to Geology*. New York: Checkmark Books.

Phillips, O.M. 1977. *The Dynamics of the Upper Ocean*, 2nd edition. Cambridge: Cambridge University Press.

Tovey, N.K. 2005. *ENV-2E02 Energy Resources 2005 Lecture*. Accessed 03/01/23 @ www2.env.ac.UK.gmmc/energy

US Department of Energy (USDOE). 2010. *Ocean Energy*. Washington, DC: Department of Interior. Accessed @ www.mms.gov

US Geological Survey (USGS). 2007. *The Water Cycle: Water Storage in Oceans*. Accessed 7/11/22 @ http://ga.water.usgs.gov/edu/watercycleoceans.html

World Energy. 2004. *Survey of Energy Sources: Wave Energy*. Accessed 03/31/23 @ www.worldenergy.org/wec-gies/publications

8 Fuel Cells

INTRODUCTION

"I believe fuel cell vehicles will finally end the hundred-year reign of the internal combustion engine as the dominant source of power for personal transportation. It's going to be a winning situation all the way around—consumers will get an efficient power source, communities will get zero emissions, and automakers will get another major business opportunity—a growth opportunity" (William C. Ford, Jr., Ford chairman, International Auto Show, January 2000, New York, New York). The point is, the fuel cell is a unique power converter that is nonpolluting, flexile, and efficient. It combines hydrogen or natural gas with oxygen via an electrochemical process to produce electricity.

FUEL CELLS: A REALISTIC VIEW

HYDROGEN[1]

Containing only one electron and one proton, hydrogen, chemical symbol H, is the simplest element on Earth. Hydrogen is a diatomic molecule—each molecule has two atoms of hydrogen (which is why pure hydrogen is commonly expressed as H_2). Although abundant on Earth as an element, hydrogen combines readily with other elements and is almost always found as part of another substance, such as water, hydrocarbons, or alcohols. Hydrogen is also found in biomass, which includes all plants and animals.

- Hydrogen is an energy carrier, not an energy source. Hydrogen can store and deliver usable energy, but it doesn't typically exist by itself in nature; it must be produced from compounds that contain it.
- Hydrogen can be produced using diverse, domestic resources including nuclear; natural gas and coal; and biomass and other renewables including solar, wind, hydroelectric, or geothermal energy. This diversity of domestic energy sources makes hydrogen a promising energy carrier and important to our nation's energy security. It is expected and desirable for hydrogen to be produced using a variety of resources and process technologies (or pathways).

          DOI: 10.1201/9781003439059-9

- The USDOE focuses on hydrogen-production technologies that result in near-zero, net greenhouse gas emissions and use renewable energy sources, nuclear energy, and coal (when combined with carbon sequestration). To ensure sufficient clean energy for our overall energy needs, energy efficiency is also important.
- Hydrogen can be produced via various process technologies, including thermal (natural gas reforming, renewable liquid and bio-oil processing, and biomass and coal gasification), electrolytic (water splitting using a variety of energy resources), and photolytic (splitting water using sunlight via biological and electrochemical materials).
- Hydrogen can be produced in large, central facilities (50–300 miles from point of use), smaller semi-central (located within 25–100 miles of use), and distributed (near or at point of use). Learn more about distributed versus centralized production.
- In order for hydrogen to be successful in the marketplace, it must be cost competitive with the available alternatives. In the light-duty vehicle transportation market, this competitive requirement means that hydrogen needs to be available untaxed at $2–$3/gge (gasoline gallon equivalent). This price would result in hydrogen fuel cell vehicles having the same cost to the consumer on a cost-per-mile-driven basis as a comparable conventional internal-combustion engine or hybrid vehicle.
- The USDOE is engaged in research and development of a variety of hydrogen production technologies. Some are further along in development than others—some can be cost competitive for the transition period (beginning in 2015), and others are considered long-term technologies (cost competitive after 2030).

Infrastructure is required to move hydrogen from the location where it's produced to the dispenser at a refueling station or a stationary power site. Infrastructure includes the pipelines, trucks, railcars, ships, and barges that deliver fuel, as well as the facilities and equipment needed to load and unload them.

Delivery technology for hydrogen infrastructure is currently available commercially, and several US companies deliver bulk hydrogen today. Some of the infrastructure is already in place because hydrogen has long been used in industrial applications, but it's not sufficient to support widespread consumer use of hydrogen as an energy carrier. Because hydrogen has a relatively low volumetric energy density, its transportation, storage, and final delivery to the point of use comprise a significant cost and result in some of the energy inefficiencies associated with using it as an energy carrier.

Options and trade-offs for hydrogen delivery from central, semi-central, and distributed production facilities to the point of use are complex. The choice of a hydrogen production strategy greatly affects the cost and method of delivery.

For example, larger, centralized facilities can produce hydrogen at relatively low costs due to economies of scale, but the delivery costs for centrally produced hydrogen are higher than the delivery costs for semi-central or distributed production options (because the point of use is farther away). In comparison, distributed production facilities have relatively low delivery costs, but the hydrogen production costs are likely to be higher—lower volume production means higher equipment costs on a per-unit-of-hydrogen basis.

Key challenges to hydrogen delivery include reducing delivery cost, increasing energy efficiency, maintaining hydrogen purity, and minimizing hydrogen leakage. Further research is needed to analyze the trade-offs between the hydrogen production options and the hydrogen delivery options taken together as a system. Building a national hydrogen delivery infrastructure is a big challenge. It will take time to develop and will likely include combinations of various technologies. Delivery infrastructure needs and resources will vary by region and type of market (e.g., urban, interstate, or rural). Infrastructure options will also evolve as the demand for hydrogen grows and as delivery technologies develop and improve.

Hydrogen Storage

Storing enough hydrogen on board a vehicle to achieve a driving range of greater than 300 miles is a significant challenge. On a weight basis, hydrogen has nearly three times the energy content of gasoline (120 MJ/kg for hydrogen versus 44 MJ/kg for gasoline). However, on a volume basis, the situation is reversed (8 MJ/liter for liquid hydrogen versus 32 MJ/liter for gasoline). On-board hydrogen storage in the range of 5–13 kg H_2 is required to encompass the full platform of light-duty vehicles.

Hydrogen can be stored in a variety of ways, but for hydrogen to be a competitive fuel for vehicles, the hydrogen vehicle must be able to travel a comparable distance to conventional hydrocarbon-fueled vehicles.

Hydrogen can be physically stored as either a gas or a liquid. Storage as a gas typically requires high-pressure tanks (5,000–10,000 psi tank pressure). Storage of hydrogen as a liquid requires cryogenic temperatures because the boiling point of hydrogen at one atmosphere pressure −252.8° C.

Hydrogen can also be stored on the surfaces of solids (by adsorption) or within solids (by absorption). In adsorption, hydrogen is attached to the surface of material either as hydrogen molecules or as hydrogen atoms. In absorption, hydrogen is dissociated into H-atoms, and then the hydrogen atoms are incorporated into the solid lattice framework.

Hydrogen storage in solids may make it possible to store large quantities of hydrogen in smaller volumes at low pressures and at temperatures close to room temperature. It is also possible to achieve volumetric storage densities greater than liquid hydrogen because the hydrogen molecule is dissociated into atomic hydrogen within the metal hydride lattice structure.

Finally, hydrogen can be stored through the reaction of hydrogen-containing materials with water (or other compounds such as alcohols). In this case, the hydrogen is effectively stored in both the material and in the water. The term "chemical hydrogen storage," or chemical hydrides, is used to describe this form of hydrogen storage. It is also possible to store hydrogen in the chemical structures of liquids and solids.

Hydrogen Fuel Cell

The fuel cell uses the chemical energy of hydrogen to cleanly and efficiently produce electricity with water and heat as by-products. Fuel cells are unique in terms of variety of their potential applications; they can provide energy for systems as large as a utility power station and as small as a laptop computer.

Fuel cells have several benefits over conventional combustion-based technologies currently used in many power plants and passenger vehicles. They produce much smaller quantities of greenhouse gases and none of the air pollutants that create smog and cause health problems. If pure hydrogen is used as a fuel, fuel cells emit only heat and water as by-products.

DID YOU KNOW?

Hydrogen fuel cell vehicles (FCVs) emit approximately the same amount of water per mile as vehicles using gasoline-powered internal-combustion engines (ICEs).

A *fuel cell* is a device that uses hydrogen (or hydrogen-rich fuel) and oxygen to create electricity by an electrochemical process. A single fuel cell consists of an electrolyte and two catalyst-coated electrodes (a porous anode and cathode). While there are different fuel cell types, all fuel cells work similarly:

- Hydrogen, or a hydrogen-rich fuel, is fed to the anode where a catalyst separates hydrogen's negatively charged electrons from positively charge ions (protons).
- At the cathode, oxygen combines with electrons and, in some cases, with species such as protons or water, resulting in water or hydroxide ions, respectively.
- For polymer electrolyte membrane and phosphoric acid fuel cells, protons move through the electrolyte to the cathode to combine with oxygen and electrons, producing water and heat.
- For alkaline, molten carbonate, and solid oxide fuel cells, negative ions travel through the electrolyte to the anode where they combine with hydrogen to generate water and electrons.
- The electrons from the anode cannot pass through the electrolyte to the positively charged cathode; they must travel around it via an electrical circuit to reach the other side of the cell. This movement of electrons is an electrical current.

THE BOTTOM LINE

Fuel cells can provide carbon-free electricity with very low emissions, fairly good efficiencies at approximately 50%, and can provide both heat and electricity.

NOTE

1   Information in this section is from USDOE, 2008, Hydrogen, Fuel Cells & Infrastructure
    Technologies Program. Accessed @ www1.eere.energy.gov/hydrogenandfuelcells/product
    ion/basics.html.

RECOMMENDED READING

National Renewable Energy Laboratory (NREL). 2009. *Ultracapacitors*. Accessed 01/29/23 @
    www.nrel.gov/vehiclesandfuels/energystorage/ultracapacitors.html?print

Part II

Practical Applications

9 Green Energy Sources and Practices

SCOPE AND STRUCTURE

As mentioned in the preface, the green energy practice is broad in scope. Thus, to facilitate understanding of the various subject areas presented and for user-friendly application as a hands-on field manual this text is presented in two separate parts. Again, Part 1 describes in detail green energy and Part 2 the nine green energy fields and practices (see Figure 9.1). These fields include wind energy, solar, geothermal, hydro-power, biofuels, recycling, green roofs, hydrogen fuel cells, and weather insulating/sealing.

Figure 9.1 shows green energy sources along with examples of Green Energy Practices.

Recall that green power is a subset of renewable energy; it represents those renewable energy resources and technologies that provide the greatest environmental benefits (this is the focus of this chapter).

Recently, one of the major new buzzwords to emerge is Green Energy Jobs. Most supporters give a thumbs-up to the concept, and many pundits profess in their authoritative manner the benefits to be derived from such practices. The standard verbiage they put forward touts the benefits of cleaning up the environment, controlling global warming and creating an entirely new sector of employment. In the current economy, it is difficult to argue against the benefits derived from Green Energy Practices and Jobs: energy conservation and decent wages, a decent career path with upward mobility, and reduction of pollution and waste to benefit the environment. On the surface, all these statements make sense. However, there is a more important reason to create and sustain Green Energy Job growth and viability. That is, when Western economies begin to rebound from single- and double-dip recessions, the bounce to be achieved will be limited by a heavy lid placed over such advances by our continuing appetite for liquid hydrocarbon fuels. The lion's share of these fuels, of course, is controlled by nations and entities unfriendly to us. Thus, it is my position that Green Practices and Jobs are not just important for the environment and for creating jobs, but also to wean us away from those who choose to be greedy, to deride, and to despise everything Western, who are controlling the oil flow from the tap. Again, as soon as the West begins to recover economically (if it ever does), the hydrocarbon-tap controllers will simply raise the price of oil—and this is the bottom line as to why

DOI: 10.1201/9781003439059-11

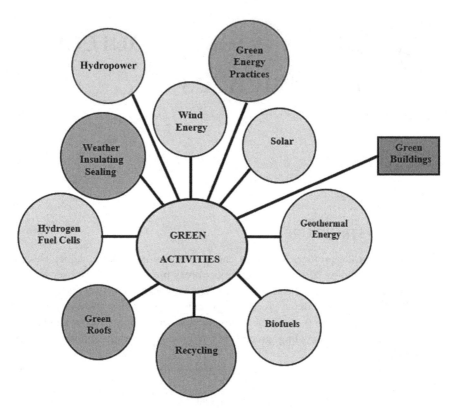

FIGURE 9.1 Green energy sources and practices.

Green Energy Practices and Jobs are important to all of us in the West. We need to shut down the foreign tap.

THE 411 ON GREEN ENERGY PRACTICES

Right up front, it is important to point out green energy practices are about conserving energy and not about producing renewable energy. OK, as shown in Figure 9.1, green energy jobs/activities (work—construction, maintenance, and operation practices) include work in wind, water, solar, geothermal, biomass, fuel cell, and marine power. For the purposes of this text, Green Energy Practices are defined as those that specifically, but not exclusively, reduce energy consumption (and rely on foreign-derived energy supplies) through high-efficiency strategies and innovation. This includes, as shown in Figure 9.1, weather insulating and sealing buildings, green roofs, and recycling. Note that other green energy practices such as Green Building protocols are available to conserve energy and prevent waste of energy.

Green Energy Practices help protect biodiversity and ecosystems; reduce materials and water consumption through high-efficiency strategies; decarbonize (a popular buzzword) the economy; and minimize or altogether avoid the generation of all forms of waste and pollution.

The evolution or "pre-history" of the Green Energy Practices/Jobs paradigm began with its main precursor, the environmental movement of the 1960s and 1970s. On its heels evolved the sustainability movement (i.e., the capacity to endure) in tandem with (parallel to; alongside) the "Green Movement," both of which bloomed in the 1990s. Since the 1980s, sustainability has become a term used in reference to environmental and human issues. Environmental, social, and economic demands are the main components that stand behind the concept of sustainability (USEPA 2011).

GREEN JOBS ACT (2007)

The Green Jobs Act of 2007 (H.R. 2847)

> authorized up to $125 million in funding to establish national and state job training programs, administered by the U.S. Department of Labor, to help address job shortages that are impairing growth in green industries, such as energy efficient buildings and construction, renewable electric power, energy efficient vehicles, and biofuels development.

> (HR 2007)

The Green Jobs Act is incorporated into the Energy Independence and Security Act that was passed in 2007.

Additionally, the American Recovery and Reinvestment Act (ARRA), passed in early 2009, makes new investments in our Nation's future. This includes creating jobs to deliver on those investments in industries including energy, as well as job training. The new focus, coupled with the move in America toward energy efficiency and more environmentally-friendly practices, is resulting in changes to traditional jobs and the creation of new kinds of occupations.

DID YOU KNOW?

Construction employs 7% of US workers, but disproportionately accounts for 21% of work-related fatalities—the largest number of fatalities reported for an industry sector.

DID YOU KNOW?

The number of back injuries in US construction was 50% higher than the average for all other US industries in 1999 (CPWR, 2002).

SIDEBAR 9.1—GREEN ENERGY: OLD PARADIGM VERSUS "NEW" PARADIGM

Let's take a moment to focus on one often overlooked element of Green Energy Practice: Environmental and Green Energy Health and Safety. The new focus on green energy production and the corresponding result of job development, coupled with the move in America toward energy efficiency and more environmentally friendly practices, is resulting in changes to traditional jobs and the creation of new kinds of occupations—the shift from the old paradigm to the so-called new paradigm in the way in which green energy workers are kept safe on the job. The point is, as we make technological advances in industry, we need to remain vigilant in protecting workers against traditional and emerging hazards. The Centers for Disease Control and Prevention (CDC, 2008) points out that as traditional jobs evolve to meet new challenges, workers may be faced with known risks that had not previously affected their occupation. These changes also present us with the opportunity to protect workers and ultimately eliminate workplace hazards.

GREEN BUILDING

Green building?

Why?

Why not? Consider that, while we know buildings and development (construction, etc.) provide countless benefits to all of us (society), they also have significant environmental and health impacts.

So, to answer these queries, let's look at some basic facts and statistics and go from there.

According to the USEPA (2016), in the United States there were 233,114 establishments/businesses in the building industry, representing more than $531 billion in annual revenues, nearly 562 billion in annual payroll, and more than 1.7 million employees in 2002 (2002 Economic Census). Residential buildings comprising a total of 128 million residential housing units existed in the United States in 2007 (USDC, 2008). Approximately 7,188 new housing units were built between 2005 and 2009 (USDC, 2007). Regarding commercial buildings, nearly 4.9 million office buildings existed in 2003 in the United States (EIA, 2003). Every year, approximately 170,000 commercial buildings are constructed, and nearly 44,000 commercial buildings demolished (1995) (USDC, 1995). Almost 84 million Americans (including 737 million students) spend their days in approximately 124,110 colleges, universities, and public and private primary and secondary schools (USDE, 2007).

Regarding energy usage, the USDOE in 2007 reported that buildings accounted for 38.9% of total US energy consumption in 2005. Residential buildings accounted for almost 54% of that total, while commercial buildings accounted for the other 47%. Buildings accounted for more than 70% of total US electricity consumption in 2006, and this number is expected to rise to 75% by 2025. Fifty-one percent of that total was attributed to residential building use, while 49% was attributed to commercial building usage. Note that the average household spends around $2,000 a year on energy bills—more than half of which goes to heating and cooling (USEPA, 2023).

Out of the total energy consumption in an average household, 50% goes to space heating, 27% to run appliances, 19% to heat water, and 4% goes for air conditioning (DOE/EIA, 2023).

Regarding air and atmosphere, the USDOE (2008) reported that buildings in the United States contribute almost 40% of the nation's total carbon dioxide emissions, including about 21% from the residential sector and 18.0% from the commercial sector (DOE/EIA, 2007).

DID YOU KNOW?

The annual mean air temperature of a city with 1 million people or more can be 1.8^0–5.4^0 F (1^0–3^0 C) warmer than its surroundings. In the evening, the difference can be as high as 22^0 F (12^0 C). Heat islands can increase summer-time peak energy demand, air conditioning costs, air pollution and greenhouse gas emissions, heat-related-illness, and mortality (USEPA, 2022). One study estimates that the heat island effect is responsible for 5%–10% of peak electricity demand for cooling buildings in cities (Akbari, 2005).

WHAT'S IT ALL ABOUT?

The practice of green building (aka as a sustainable or high-performance building) is all about creating structures using processes that are environmentally responsible and resource efficient throughout a building's life cycle—from siting to design, construction, operation, maintenance, renovation, and deconstruction—is what the 'practice' is all about. What this practice does is expand and complement the classical building design concerns of economy, utility, durability, and comfort.

So, what is the impact of the built environment? It begins with the aspects of siting, design, construction, operation, maintenance, renovation, and deconstruction. The built environment consumes energy, water, materials, and natural resources. Environmental effects of a built environment include waste, air pollution, water pollution, indoor pollution, heat islands, stormwater runoff, and noise. Ultimate effects of the built environment include harm to human health, the environment, degradation, and loss of resources.

Green buildings are designed to reduce the overall impact of the built environment on human health and the natural environment by:

- efficiently using energy, water, and other resources
- protecting occupant health and improving employee productivity
- reducing waste, pollution, and environmental degradation

For example, green buildings may include sustainable material in their construction, (e.g., reused, recycled content, or made from renewable resources); create healthy environments with slight pollutants (e.g., decreased product emissions); and/or feature landscaping that reduces water usage (e.g., by using native plants that survive without extra watering) (USEPA, 2016).

THE BOTTOM LINE

The built environment has a vast impact on the natural environment, human health, and the economy. By adopting green building strategies, we can maximize both economic and environmental performance. Green construction methods can be integrated into buildings at any stage, from design and construction to renovation and deconstruction. However, the most significant benefits can be obtained if the design and construction team takes an integrated approach from the earliest stages of a building project.

REFERENCES AND RECOMMENDED READING

2002 Economic Census. *Census Bureau*, US Department of Commerce. Accessed 7/4/23 @ www.census.gov/econ/census02/advance/TABLE2.HTM
Akbari, H. (2005). *Energy Saving Potentials and Air Quality Benefits of Urban Heat Island Mitigation*. Berkeley, CA: Lawrence Berkeley National Laboratory.
Board of Certified Safety Professionals (BCSP). 2006. *Comprehensive Practice Self-Assessment Examination*, 4th edition (rev.). Savoy, IL: Board of Certified Safety Professionals.
Center to Protect Workers' Rights (CPWR). 2002. *Construction Chart Book*, 2nd edition. Silver Spring, MD: Center to Protect Workers' Rights.
Centers for Disease Control and Prevention (CDC). 2008. *Preventing Work-Related Hazards through Design*. Accessed 10/30/11 @ www.cdc.gov/features/preventiondesign/
US Department of Commerce (USDC). 1995. *C-Series Reports. Manufacturing and Construction Division, Census Bureau*. Washington, DC: US Department of Commerce.
US Department of Commerce (USDC). 2008. *American Housing Survey for the United States—2007*. Washington, DC: US Department of Commerce.
US Department of Education (USDE). 2007. *National Center for Educational Statistics-Digest*. Washington, DC: US Department of Education.
US Department of Energy (USDOE). 2008. Buildings Energy Databook, 2006. US Department of Energy and Annual Energy Review 2007. EOE/EIA-0384. Energy Information Administration. Washington, DC: US Department of Energy.
US Department of Energy/Energy Information Administration (USDOE/EIA). 2007. *Emission of Greenhouse Gases in the United States, 2007*. Washington, DC: Energy Information Administration. US Department of Energy.
US Department of Energy/Energy Information Administration (USDOE/EIA). 2023. *Changes in Energy Usage in Residential Housing Units*. Accessed 7/5/23 @ http://eia.doe.gov/emeu/recs/recs97/decade.html#toteons4
US Energy Information Administration (EIA). 2003. *2003 Commercial Buildings Energy Consumption Survey—Overview of Commercial Buildings Characteristics*. Accessed 7/4/23 @ www.eia.doe.gov/emeu/cbecs2003/introduction.html
US Environmental Protection Agency (USEPA). 2011. *EPA Sustainability*. Accessed 10/27/23 @ www.epa.gov
US Environmental Protection Agency (USEPA). 2016. *Green Building*. Accessed 7/4/23 @ https://archive.epa.gov/greenbuildig/web/html.about.html

US Environmental Protection Agency (USEPA). 2022. *Heat Island Program*. Accessed 7/6/23 @ www.epa.gov/heatisland

US Environmental Protection Agency (USEPA). 2023. *US EPA Energy Star Program*. Accessed 7/5/23 @ http://energystar.gov/index.cfm?c=thermostats.pr_thermostats

US House of Representatives. 2007. *House Committee Pass Solis' Green Jobs Act*. Accessed 10/27/11 @ http://solis.house.gov/list/press/ca32_solis/wida6/green jobscomm.shtml), US House of Representatives.

Part III

Green Infrastructure

10 Living with Nature

INTRODUCTION

Those living in the United States and paying attention to the environmental impacts of climate change see almost daily the occurrence of increasingly severe storms that are growing in intensity and frequency, resulting in disruptive and urban flooding. These flooding events are associated with property loss, displacement, lost wages, economic disruption, and physical health issues. Without a doubt, these issues with their associated impacts exacerbate existing community challenges including, for example, stormwater runoff, which continues to be a major cause of water pollution in urban areas. It carries pollutants such as trash, bacteria, and heavy metals through storm sewers into local waterways. Heavy rainstorms can cause flooding that damages property and infrastructure.

In the past, it was common practice to use gray infrastructure—systems of gutters, pipes, and tunnels—to move stormwater away from where we live to treatment plants or directly to local water bodies. Note that gray infrastructure and practices have not gone away. A major problem with gray infrastructure is it's aging. Moreover, its existing capacity to manage large volumes of stormwater is decreasing in areas across the United States. Because of this problem, many communities are installing green infrastructure systems to bolster their capacity to manage stormwater. By doing so, communities are becoming more resilient and achieving environmental, social, and economic benefits.

What green infrastructure is all about is collecting and filtering stormwater where it falls. In 2019, Congress enacted the Water Infrastructure Improvement Act, which defines Green Infrastructure as

> the range of measures that used plant or soil systems, permeable pavement or other permeable surfaces or substations, stormwater harvest and reuse, or landscaping to store, infiltrate, or evapotranspirate stormwater and reduce flows to service system or to surface waters.

> (USEPA, 2023)

Note that green infrastructure elements can be woven into a community at several scales. Examples on the urban scale could include a rain barrel up against a house, a row of trees along a major city street, or greening of an alleyway. Neighborhood-scale green infrastructure could include acres of open park outside a city center, the planting of rain gardens, or construction of a wetland near a residential housing complex. At the watershed scale, examples could include protecting large, open natural spaces. When green infrastructure systems are installed throughout a community, city, or across a regional watershed, they can provide cleaner air and water as well as significant value for the community with flood protection, diverse habitat, and beautiful green spaces (USEPA, 2023).

DID YOU KNOW?

There is nothing new about Green Infrastructure practices. They can be traced as far back as the seventeenth century in European society beginning in France (Jones, 2021). France used the presence of nature—living with nature—to provide social and spatial organization to their towns (Jones, 2021).

TYPES OF GREEN INFRASTRUCTURE

Presently, based on the author's experience, there are 11 types of Green Infrastructure:

- Green Roofs
- Downspout Disconnection
- Rainwater Harvesting
- Rain Gardens
- Planter Boxes
- Bioswales
- Permeable Pavements
- Green Parking
- Land Conservation
- Green Streets and Alleys
- Urban Forests (Urban Tree Canopy)

GREEN ROOFS

1856–1927 Time frame
Place: South Dakota Prairie, USA
Subject: "Nebraska Marble"
There were trees, no nails, no bricks, no rock. What they needed to begin with was safe, secure, livable shelter in the South Dakota prairie. When the Homestead Act was passed by the US Congress in 1862, there was a steady migration of hundreds of thousands of intrepid souls who wagoned their way to the western prairies to homestead 160 acres of free land.

So, with conventional-type building materials such as trees, nails, bricks, and rocks nonexistent, and tents and the top of a covered wagon providing little safety, security, and relatively comfortable living, what did the pioneer do to provide housing/shelter other than a tent or a prairie schooner top?

The only building material available: prairie sod, jokingly called "Nebraska marble." Sod is the top layer of earth that includes grass, its roots, and the soil clinging to the roots. Building a sod house (aka soddy) during the sunburn of the day or during the white of a moonlit night was no easy task—just lots of work and often taking many weeks, especially if the settlers had no neighbors nearby to help. And because of the prairies' wild weather, many settlers started building dugouts. These small, dark spaces were dug into the side of a hill (i.e., if a hill were included in the 160-acre homestead) that could be made quickly, and were drier and warmer than tents. Later, many settlers built a sod house right in front of the dugout and utilized the dugout as an extra room.

There was no easy way to cut sod. There were no D-9 bulldozers, backhoes, or other earth-digging or -moving equipment. Simply put, cutting sod was no picnic without using motorized tractors because they were not available until later in the 1910s. So, what the settlers used instead were oxen, horses, or mules, and special plows equipped with curved steel blades to cut through the sod. Experience confirmed that the settlers could only cut enough sod for one-day use. The problem was that sod dried without delay and cracked and crumbled if not used right away. Farmers cut sod close to the location where they wanted to build their soddy. The benefit of building a soddy for shelter was that it protected the inhabitants from prairie fires. Moreover, removing the grass from the area worked to keep insects, snakes, and other vermin from burrowing into the soddy.

It took about 3,000 bricks of sod, that were 8 inches wide by 24 inches long, weighing close to 50 pounds each, to build a 16-by-20-foot house. The freshly cut sod bricks were stacked root-side up in order to marry the roots to the brick above it. This was a smart move because eventually the sod bricks grew together to form a solid wall—a very strong wall. These soddies required a wide foundation. Also, the walls sloped downward on the outside of the soddy so that as the sod walls settled, they were not subject to collapse. The odd thing was that the top of the soddy looked smaller than the bottom foundation.

Anyway, let's get a back to green roofs.

Putting a roof on a soddy was not easy and was dangerous, too. Because there was a lack of normal roof materials, like slate tiles, or wooden shingles, there was a need to produce alternative materials such as mud, grass, and sod. A series of cedar poles (if available) were used to hold up layers of brush tied into bundles, coated with mud or made up of sod and grass. Not the greatest construction technique but necessary in the prairie.

The point of bringing the construction of prairie shoddies into this discussion is simple: There is nothing new about green roofs. There is a difference, however, in their purpose. The difference and purpose today? Well, in the day of prairie soddy construction, what we today call a green roof, a vegetated roof, or an eco-roof had only one environmental goal at that time and that was to provide a roof over the soddy and thus protect the inhabitants' presence in a roofed and walled shelter. However, storms and heavy rains tended to do damage to the soddy. Moreover, snakes also kind of liked the sod.

DID YOU KNOW?

A lesser-known cousin of a green roof, the blue roof, is a design that is not technically a green infrastructure. What it does is collect and store rainfall, reducing the inrush of runoff water into sewer systems. Blue roofs used detention basins or detention ponds, for collecting rainfall before it gets drained into waterways and sewers at a controlled rate.

LIVING ROOFS

Also known as "living roofs," green roofs are steadily growing in urban areas throughout the United States. And as pointed out, green roofs have been around for a long time. Prairie homesteaders built soddies when settling the frontier. Green roof projects serve several purposes for a building, such as absorbing rainwater, providing insulation, creating a habitat for wildlife, and helping to lower urban air temperatures. Many of these projects are installed by companies that specialize in green roof technology, or more traditional contractors like landscapers, roofing contractors, or the amalgamation of these traditional trades, can also be involved in the installation of green roofs. In the day of the frontier soddy, when the soddy was ready to be roofed, this was a dangerous operation. Note that today the installation and ongoing maintenance of a green roof falls under Federal OSHA's general industry standards. The safety challenges associated with this growing industry have new and very often familiar safety issues. Workers in the green roofs industry are exposed to typical workplace hazards.

Today, living roofs consist of a waterproofing membrane, soil (growing medium), and plants (vegetation) overlying a traditional roof in a city environment. The purpose of these living roofs is to help mitigate the problems cities create by bringing natural cooling, water treatment, and air filtration properties that vegetated landscapes provide to the urban environment. Bringing nature back to the city is what living roofs are all about. Architects and planners use living roofs to help solve environmental problems. Living roofs are an important tool to increase sustainability and biodiversity, and (the key point) they decrease energy consumption.

The truth be told, there have been living roofs on US government buildings and parking structures for almost a century—living green roofs installed on several federal buildings in the Washington, DC, region have not been replaced since their installation in the 1930s. While it is true that there have not been many for decades, living green roofs are now being revived and studied for their environmental benefits. The General Services Administration (GSA) points out that the growth of living roofs in the United States mirrors their use in other countries, like Germany, where they are more commonly seen. Moreover, the GSA has designed and maintained green roofs for decades and finds them to be economical amenities that make fiscal and environmental sense. In 2011, the GSA maintained at least 24 living roofs in 13 cities around the country. The GSA continues to install living green roofs on new and existing buildings.

TALK ABOUT ROOFS

It is a good time in this discussion to talk about roofs, in general. The roofs that we are most likely familiar with are conventional roofs—otherwise known as *black roofs*, their traditional color. Black roofs are descendants of the "tar beach" roofs once common in urban areas. Pollak (2004, p. 2) stated that black beach roofs are where you get "the sun but no sand—it is the urban alternative to the Hamptons." Keep in mind that these black beach roofs were composed of petroleum-based materials. Whether composed of polymer-modified material or synthetic rubber, these black roofs absorb significant energy from the sun and can reach temperatures higher than 150^0 F in summer. Runoff from black roofs, during storms, flows immediately into storm-wastewater (sewer) systems, contributing to flooding and water pollution. Obviously, this creates environmental and human health hazards, particularly where stormwater and wastewater systems are combined. Keep in mind that excessive stormwater runoff erodes soil and riverbanks and adds sediments and street pollutants to water bodies.

Not all roofs are living roofs or black beach tar roofs. A white roof (aka cool roof) is also commonly used in buildings and is growing in popularity. These roofs are made of light-colored material and do not heat up as much as black roofs in the sun. Note that white roofs share the same runoff problems as the black roofs.

TYPES OF GREEN ROOFS

A green roof construction project includes a waterproof barrier to protect the structure, a drainage layer to store and direct runoff, a soil or growth medium layer, and a plant layer. There are two types of green roofs: *extensive roofs*, which are relatively inexpensive to install and are used mainly for environmental benefit, and *intensive roofs*, which allow a greater variety and size of plants such as shrubs and small trees but that are usually more expensive to install and maintain, partly due to the need for irrigation. Usually, commercial and public buildings tend to use extensive roofs unless the roofs are intended primarily as occupied garden featured space.

Extensive roofs have a thin soil layer and feature succulent plants like sedums that can survive in harsh conditions. These plants have low growth height, rapid growth/spreading, and fibrous roots that have high drought tolerance and minimize water loss. An intensive green roof is more like a conventional garden or park and can have virtually no limit to plant types that can be used because of deeper plantings. The limiting factor in intensive roof construction is weight; the roof must be designed to accommodate heavier planting medium and larger root systems. These plants may require irrigation during dry periods.

GREEN ROOF BENEFITS

The benefits of installing green roofs on commercial and public buildings include (GSA, 2011):

- reduced energy consumption in some environments
- improved water quality due to reduced stormwater runoff and fewer overflows of combined sanitary and stormwater wastewater systems
- increased habitat promoting biodiversity
- improved air quality
- improved sound absorption in the top floors of buildings
- lower temperatures for building roofs and the air above them in most climates

Case Study 10.1—Green Roofs in Kansas City[1]

Kansas City is continually adding more green roofs to its city skylines. Because of city leaders' commitment to green infrastructure and their desire to improve local water and air quality, the USEPA worked with the Mid-America Region Council (MARC)—the metropolitan planning organization of the Kansas City metro area—to create a study to quantify the numerous benefits of green roofs that address various environmental challenges. This group of environmental professionals, landscape architects, business developers, and city planners highlighted the need to preserve and protect drinking water, air quality, and ecosystems.

In 2016, MARC partnered with the various environmental organizations to host a Green Infrastructure Charrette. A group of 60 professionals and community members participate in a workshop to prioritize green infrastructure goals, build local partnerships, and commit to implementing key strategies that connect citizens of Kansas City with the natural surrounding environment. The goal of this group is to preserve and protect drinking water, air quality, and ecosystems.

WATER QUALITY AND STORMWATER MANAGEMENT

One of the major targets of the Kansas City planning group is to make improvements to reduce its 54 billion gallons of annual combined sewer overflow discharges to comply with the Clean Water Act. These infrastructure improvements include a commitment to use green infrastructure to reduce stormwater runoff into the city's combined sewer system. Since 2002, the city has discharged more than 6.4 billion gallons of untreated sewage each year into local streams, rivers, and their tributaries, including the Missouri River, Fishing River, Blue River, Brush Creek, Todd Creek, and Penn Valley Lake. Sewer overflows can pose risks to public health and sanitation, as well as damage city and private property. So, city planners and city managers began to incorporate green roofs to help control and prevent stormwater runoff that contributes to these overflows along with other mitigation measures.

URBAN HEAT ISLANDS

Elevated temperatures from heat islands can affect a community's environment and quality of life in several ways. Heat islands increase demand for air conditioning to cool buildings. The USEPA (2023) reports that in an assessment of case studies spanning locations in several countries, electricity demand for air conditioning increased approximately 1%–9% for each 2^0 F increase in temperature. Santamouris (2020) points out that in countries where most buildings have air conditioning, such as the United States, had the highest increase in electric demand. This increase in demand, of course, contributes to higher electricity expenses.

Well, we know that heat islands increase both electricity demand, as well as peak energy demand. Zamuda et al. (2018) point out that it is the hot summer weekday afternoons, when offices and homes are running air-conditioning systems, lights, and appliances, when peak demand occurs. During extreme heat events, which are exacerbated by heat islands, the increased demand for air conditioning can overload systems and require a utility to institute controlled, rolling brownouts or blackouts to avoid power outages.

As described above, elevated emissions of air pollutants and greenhouse gases from heat islands raise demand for electricity. The truth be told, it is the companies that supply electricity which typically rely on fossil fuel power plants to meet much of the demand, that in turn leads to an increase in air pollutants and greenhouse gas emissions.

These pollutants are nothing to laugh about or ignore, because they are harmful to human health and contribute to complex air quality problems such as the formation of ground-level ozone (smog), fine particulate matter, and acid rain. Increased use of fossil-fuel-powered plants also increases emissions of greenhouse gases, such as carbon dioxide, which contributes to global climate change.

Elevated temperatures can directly increase the rate of ground-level ozone formation and impact energy-related emissions. Ground-level ozone is formed when nitrogen oxides and volatile organic compounds react in the presence of sunlight and hot weather. If all other variables are equal, such as the level of precursor emissions in the air and wind speed and direction, more ground-level ozone will form as the environment becomes sunnier and hotter.

In 2014, Kenward et al. reported that Kansas City was ranked one of the top US metro areas that experiences intense summer urban heat islands—measured as the greatest difference between average temperatures in rural and urban areas over an entire summer. The Kansas City metro area was 4.6^0 F warmer on average than the surrounding rural area during summer months—ranking it as the seventh-largest differential in the United States.

The bottom line: green roofs can help alleviate the urban heat island effect.

DOWNSPOUT DISCONNECTION

The USEPA (2020) points out that downspout disconnection is a form of green infra-structure that separates roof downspouts from the wastewater (sewer) system and redirects roof water runoff into permeable surfaces. This system is especially benefi-cial to cities with combined sewer systems. During a heavy rainstorm, every down-spout on typical homes can send 12 gallons of water a minute to the wastewater system, which increases the risk of wastewater (raw sewage) backing up into basements and overflowing into waterways. This simple practice drains rainwater into rain barrels (see Figure 10.1), cisterns, ground-level pits, aquifers, or permeable areas. This system can store stormwater for usage such as watering grass, flower gar-dens, and vegetable gardens.

RAINWATER HARVESTING

As a conservation measure, there is nothing new about rainwater harvesting; it has been used throughout history, particularly in regions where other water resources are scarce or difficult to access. As a part of green infrastructure, rainwater harvesting systems reduce stormwater pollution by slowing runoff and collecting rainfall for future use. The variety of systems range from the backyard rain barrel (see Figures 10.1 and 10.2) and the commercial building cistern to ground-level pits and aquifers.

FIGURE 10.1 Downspout disconnection and reconnection to rain barrel. Photo by F. Spellman.

FIGURE 10.2   Rain barrel used to water boxed plants. Photo by F. Spellman.

RAIN GARDENS

Rain gardens (aka bioretention practices) are landscaped depressions—small, shallow, sunken areas of plantings—that treat on-site stormwater discharge from impervious surfaces such as roofs, driveways, sidewalks, parking lots, and compacted lawns. The designs of rain garden practice mimic volume reduction and pollutant removal mechanisms that work in natural systems. The filtered stormwater soaks into the ground, provides water to plants, and can help recharge the local aquifers. Through these processes, rain gardens reduce peak flows within downstream wastewater systems and allow pollutant removal through filtration and plant uptake. Note that rain gardens are not suitable for treating large drainage areas. The problem is that surface soil layers can clog over time in areas with excessive sediment loadings.

The bottom line: rain gardens function as designed but require landscape maintenance as well as measures to ensure that the practice does function.

PLANTER BOXES

Before the 9/11 terrorist attacks on the United States, it was quite common to see bollards and sections of Jersey Wall used to provide a perimeter in front of governmental agencies, private businesses, and public spaces, to protect people from vehicle ramming attacks. They are also commonly used to protect utilities, electronics, machinery, and some are also used as a visual guide for vehicular traffic and to mark boundaries. After 9/11, security professionals opted for using barriers of the

FIGURE 10.3   Planter boxes. Photo by F. Spellman.

type that are more robust and weightier in their construction and protective value than bollards and Jersey Wall sections, although these two protective barriers are still in common use today. So, the trend after 9/11 and other domestic uprisings, and criminal or terrorist acts is to use heavy-duty and weightier planter boxes instead of bollards and Jersey Wall sections. Today, it is common to observe planter boxes in the place of bollards and Jersey Wall sections to provide a more robust and weightier barrier to protect city and government properties.

Regarding planter boxes (see Figures 10.3 and 10.4) used for security and protective purposes, it is also important to point out that they are a key element in green practices; these concrete planters and flowerpots also serve to aid the environment. Plant boxes used in urban settings are basically urban rain gardens with vertical walls and either open or closed bottoms. The point is these plant boxes, usually found in downtown areas, collect and absorb runoff from streets, sidewalks, and parking lots. They are ideal for areas with limited space and are a useful way to beautify city streets.

BIOSWALES

Another common green practice is the use of modified geometric bioswales that are essentially rain gardens placed in long narrow spaces such as the space between the sidewalk and the curb, or between curb and roadway (see Figures 10.5 and 10.6). In the context of stormwater management, to improve water quality, a grassed swale is a vegetated, open-channel green management practice that treats and reduces

FIGURE 10.4   Other plant containers. Photo by F. Spellman.

FIGURE 10.5   Bioswale.

FIGURE 10.6   Another view of bioswale.

stormwater flows for specified water quality volume. As stormwater flows along these swales, the vegetation slows it down, allowing for sedimentation, soil filtration, and/ or infiltration into the underlying soils. Grassed swale includes the grassed channel, dry swale, bioswale, and wet swale—all of these are improvements on the traditional drainage ditch. Basically, bioswales are used as a treatment technique and conveyance practice.

DID YOU KNOW?

Swales are linear practices, well suited for treating stormwater from roadways or residential roads and trails. Bioswales are an important part of a treatment train that includes, for instance, conveying water to a rain garden and receiving water from filter strips. Moreover, swales can be integral parts of green infrastructure and better site design approaches. Better site design is a design strategy that aims to reduce impervious cover, preserve natural lands, and capture stormwater on-site (USEPA, 2021).

PERMEABLE PAVEMENTS

Permeable pavements are a green infrastructure stormwater control that allows stormwater to infiltrate through the surface of the pavement to the ground below—a green infrastructure alternative to traditional impervious surfaces. Types of permeable pavements and walkways include porous asphalt, pervious concrete, and permeable interlocking pavement (PICP).

Regarding asphalt (aka pervious, permeable, popcorn, or open-grade asphalt) and pervious concrete (aka porous, gap-graded, or enhanced porosity concrete), these are versions of traditional asphalt or concrete with reduced sand and fines to allow for greater porosity and infiltration. PICP consists of pavers (aka manufactured concrete units) with small openings between permeable joints that contain highly permeable, small-sized aggregates.

The bottom line: with proper installation, permeable pavements can serve as durable, low-maintenance, and low-cost alternatives to traditional impermeable pavements.

GREEN PARKING

The term "Green Parking" refers to several techniques that together, reduce stormwater discharge from parking lots. Green parking techniques include setting the maximum number of parking spaces, right-sizing the dimensions of parking spaces, substituting alternative surfaces for asphalt in overflow parking areas, using green infrastructure to treat stormwater, encouraging shared parking, and providing economic incentives for planned parking.

The bottom line: many green infrastructure elements can be seamlessly incorporated into parking lot designs. Permeable pavements can be installed in sections of a lot, and rain gardens and bioswales can be included in medians and along the parking lot perimeter. When built into a parking lot, these elements also reduce the heat island effect and improve walkability in the area.

LAND CONSERVATION

Green infrastructure that includes land conservation planning and projects makes good sense for local community planners and officials to implement to protect air and water, wilderness and wildlife, and to protect the inhabitants of such areas. Local officials who concern themselves with the negative impacts of suburbanization on natural systems, agriculture, and quality of life are on the right track in efforts to protect our way of life. Experience has indicated to these far-thinking planners and officials that the loss of forests, farms, and natural areas, and past land development practices have been linked to water and air quality degradation. Moreover, these far-thinking planners and officials understand and anticipate that accelerated population growth is inevitable and that they must be able to foresee future environmental impacts of "too" many people—or at least the impact on green infrastructure as a result of growing numbers of people, which, for example, to accommodate the increased populations, may result in the removal of natural areas for planned communities that could result in environmental degradation issues, for example, from sea-level rise in natural areas.

The movement of waves of people from major cities to other areas (states) to escape dangerous weather conditions (e.g., tornadoes, droughts, historic floods, wildfires, water quality issues, etc.), poor economic opportunities, increases in drug abuse, homelessness, crime, and overall degradation of quality of life in general is occurring, or is likely to occur whenever accelerated population growth results in the modification of natural areas.

The bottom line: whenever natural areas are converted to accommodate human population growth, systems that use natural processes which result in the natural infiltration, evapotranspiration, or use of stormwater to protect water quality and associated aquatic habitat are degraded to the point where recovery to preexisting environmentally friendly areas is difficult if not impossible.

GREEN STREETS AND ALLEYS

Using plants, soils, landscape design, and engineered techniques to retain, absorb, and reduce polluted stormwater runoff is the goal of integrating green infrastructure elements into street and alley design. Green streets and alleys are designed to store, infiltrate, and evaporate and transpire stormwater while adding aesthetics to landscapes. Green streets and alleys prevent or reduce the amount of runoff that flows directly into storm drains and can be a vital tool for cities to address combined sewer overflows and nutrient impairment. Green streets and alleys can be designed as open spaces to allow light to mitigate heat island effect and to capture stormwater and make the streets and alleys usable by and for the neighboring communities (Green Alleys, 2018).

URBAN TREE CANOPY

OK, before discussing urban tree canopies, it is important to talk about what is a forest land.

Forest Land

Land at least 10 percent stocked by forest trees of any size, including land that formerly had such tree cover and that will be naturally or artificially regenerated. Forest land includes transition zones, such as areas between heavy forested and nonforested lands that are at least 10 percent stocked with forest trees and forest areas next to urban and built-up lands. Also included are pinyon-juniper and chaparral areas in the West and afforested areas. The minimum area for classification of forest land is 1 acre. Roadside, streamside, and shelterbelt strips of trees much have a crown width of at least 120 feet to qualify as forest land. Unimproved roads and trails, streams, and clearings in the forest areas classified as forest if they are less than 120 feet wide.

(Spellman, 2023, p. 23)

Note that what is described above is the type of forests—natural forest lands—that a lot of people (city types especially) call the "woods"—"let's go camping in the woods" has commonly been stated by those who want to escape urban life, even for a short time. This is the old "let's get one with nature."

Wanting to get back to nature is a noble craving—looking for peace, peace of mind, the calmness that only natural settings provide. It is these "calming" natural settings that urban forestry is all about.

Note that millions of acres of America's forests are located right in our cities and towns. These forest areas come in many different shapes and sizes. They include urban parks, street trees, landscaped boulevards, gardens, river and coastal promenades, river corridors (riparian areas), greenways, nature preserves, shelter belts of trees, and working sites at former industrial complexes. Through planned connections of green spaces, urban forests form the green infrastructure on which communities depend.

Urban populations are growing and outpacing the national growth average. What this means is that we are becoming an urbanized population. Because of this growth pattern, urban forests are more important than ever—they are the trees outside our homes. These urban forests help filter air and water, aid in stormwater management, conserve energy, reduce noise, and provide shade and habitat for animals. Urban design can also be important to residents because urban forests not only provide form but also beauty to our communities.

THE BOTTOM LINE

Scientists have been warning us of the catastrophic harm that can be done to the world by atmospheric warming. Various observers state that the effect could bring record droughts, record heat waves, record smog levels, and an increasing number of forest fires.

Another caution put forward warns that the increasing atmospheric heat could melt the world's ice caps and glaciers, causing ocean levels to rise to the point where some low-lying island countries would disappear, while the coastlines of other nations would be drastically altered for ages—or perhaps for all time.

What's going on? We hear plenty of theories put forward by doomsayers, but are they correct? If they are correct, what does it all mean? Does anyone really know the answers? Should we be concerned? Should we invest in waterfront property in Antarctica? Should we panic?

No. While no one really knows the answers—"we don't know what we don't know syndrome"—and while we should be concerned, no real cause for panic exists.

Should we take some type of decisive action—should we produce quick answers and put together a plan to fix these problems? What really needs to be done? What can we do? Is there anything we can do?

The key question to answer here is "What really needs to be done?" We can study the facts, the issues, the possible consequences—but the key to successfully combating these issues is to stop and seriously evaluate the problems. We need to let scientific facts, common sense and cool-headedness prevail. Shooting from the hip is not called for, makes little sense—and could have Titanic consequences for us all.

The other question that has merit here is, "Will we take the correct actions before it is too late?" The key words here are: "correct actions." Eventually, we may have to take some action (beyond hiding in a cave somewhere). But we do not yet know what those actions could be or should be.

From our perspective, one thing is certain: in our college-level environmental health courses, we address, sooner or later, global warming and/or global climate change. Through time and experience we have learned (yes, teachers learn, too) that whether we call it global warming, global climate change (humankind-induced global warming, under a broader label), or an inconvenient truth, the topic is a conundrum (riddle, the answer of which is a pun). As such, before diving into the many emotionally charged, heated class discussion about this "hot" topic (pun intended), we are reminded by two celebrated statements of just how complicated a conundrum can be.

Consider this: any damage we do to our atmosphere affects the other three environmental mediums: water and soil—and biota (us—all living things). Thus, the endangered atmosphere (if it is endangered) is a major concern (a life and death concern) to all of us.

NOTE

1 From USEPA 2018. Estimating the Environmental Effects of Green Roofs: A Case Study in Kansas City, Missouri. Washington, DC: United States Environmental Protection Agency.

REFERENCES AND RECOMMENDED READING

Green Alleys. 2018. *The Trust of Public Land*. Accessed 7/16/23 @ www.tpl.or/green-all eys#sm.000001w1n9rvfo8u6y3y8zmk5xa

Jones, K.R. 2021. Greening the city: Nature in French Towns from the 17th century. *Environment and History* **27**(4): 9–10.

Pollack, M. 2004. Sun but no sand. New York Times, Section 14, p. 2.

Santamouris, M. 2020. Recent progress on urban overheating and heat island research. *Energy and Buildings* **207**: 109482.

Spellman, F.R. 2023. *Science of Carbon Sequestration and Storage*. Boca Raton, FL: CRC Press.

US Environmental Protection Agency (USEPA). 2020. *What Is Green Infrastructure?* Accessed 7/16/23 @ www.epa.gov/green-infrastructure/what-green-instructure

US Environmental Protection Agency (USEPA). 2021. *Stormwater Best Management Practice Grassed Swales*. Accessed 6/6/23 @ wwwl.epa.gov/system/files/documents/2021-11/bmp-grassed-swales.pdf

US Environmental Protection Agency (USEPA). 2023. *Green Infrastructure*. Accessed 7/10/23 @ www.epa.gov/green-infrastructure

US General Services Administration (GSA). 2011. *The Benefits and Challenges of Green Roofs on Public and Commercial Buildings*. Washington, DC: General Services Administration.

Zamuda, C., Bilello, D.E., Conzelmann, G., Mecray, E., Satsangi, A., Tidwell, V., and Walker, B.J. 2018. *Energy Supply, Delivery, and Demand*. Washington, DC: US Global Change Research Program, p. 174–201.

11 Weather Insulation and Sealing

INTRODUCTION

Weatherizing, insulating, and sealing are important tools in the green energy toolbox. They are not only used to increase energy efficiency and conservation programs but also create a market for green jobs. Various types of weatherizing, insulating, and sealing products include blow-in and spray-on applications. These products can help reduce energy consumption by improving the insulation of buildings and homes. This can help reduce the amount of energy needed to heat or cool a building or home, which can help reduce greenhouse gas emissions. Green infrastructure can also help reduce the amount of energy needed to manage water by reducing rainwater flows into sewer systems. This can help reduce pumping and treatment demands for municipalities.

WEATHERIZING, INSULATING, AND SEALING

The USEPA (2009) points out that sealing and insulating the "envelope" or "shell" of your home—its outer walls, ceilings, windows, doors, and floors—is often the most cost-effective way to improve energy efficiency and comfort. Weatherizing, insulating, and sealing are important parts of green energy efficiency and conservation programs, creating a market for green jobs. Various types of weatherizing, insulating, and sealing products include blow-in and spray-on applications. These may require a certain amount of training to apply these materials to product specifications so as to achieve proper weatherization. The materials used in these types of applications can include, but are not limited to blow-in fiberglass and cellulose; and for spray-on, spray polyurethane foam (SPF), spray polystyrene, and spray latex sealant.

CREATING THE ENVELOPE: A PERSONAL EXAMPLE

Note to Reader: The best way to explain the green infrastructure practice of creating an envelope around your house or other structure is to provide the following personal example on how to accomplish this, utilizing the USEPA's Energy Star Program® as the blueprint, the game plan. In the following I explain how to seal and insult a 2,500-square-foot two-story home by following the directions provided by the Energy Star

Program—the government-backed green program that aims to help us save money and protect our environment.

xxxxxxxxxxxxxxxxxxxxxxxxxxxxxxxxxxxxxxxxxxxxxxxxxxxxxxxxxxxxxxxxxxxxxx

Note that the USEPA (2009) points out that the green practice of "Sealing and Insulating" your home is all about creating a "shell" or, better yet, an "envelope" to surround and protect your living spaces. To accomplish this green practice, to create the envelope, the shell or envelope is accomplished by sealing outer walls, ceilings, windows, doors, and floors, which in turn conserves energy, increases energy efficiency, provides comfort, and enhances well-being. Also, reducing the use of energy, from any source, is cost effective.

In total, the benefits gained by sealing and insulating structures include:

- Reduced energy costs
- Improved comfort
- Reduced noise from outdoors
- Reduced levels of dust, pollen, and insects entering homes
- Much better humidity control

STEP 1

So, for my first step in the sealing and insulating of my 2,500-square-foot home, I knew I needed a plan. Of course, I could have hired a planner and construction experts to seal and insulate my home. Appropriately trained contractors would have little difficulty in enveloping my house, so to speak, but possessing basic skills in carpentry, plumbing, and electrical work, and wanting to economize, I decided to do the project myself. Again, I could have hired a contractor to do the work; however, I had economizing in mind, so I decided to perform do-it-me insulating and sealing.

If your plan is like mine, a self-help, do-it-me project, then a prudent first step is to follow the USEPA's Energy Star Program®, a government-backed program designed to help individuals or businesses seal and insulate to ensure energy efficiency where cost effective. The Energy Star program is designed to improve the envelope of a structure in order to make it not only more comfortable but also energy efficient.

My first step was to study the USEPA's Energy Star written guidance. After reading the guide, I followed the instructions to a T. First, I sketched my home's floor plan. Basically, this rough sketch (rough because I am not a skilled draftsperson) served as a reference point for when I got into the attic and helped me locate areas of leakage. Note that when you draw your sketch, it should include home or a building's dropped soffits over kitchen cabinets or bath vanities, slanted ceilings over stairways, where interior or exterior walls meet the ceiling, and any other dropped ceiling areas. By the way, a soffit, by definition, is the underside of a building overhang beam or arch, especially the underside of a stair or roof overhang. These areas may have open stud cavities leading directly into the attic and can be huge sources of air leaks.

I can't overemphasize the need to start with a plan. The key to any successful home improvement project is adequate planning. Also, as part of the plan, you need to gather the tools and supplies before you begin to minimize trips in and out of the attic. The first item needed in order for you to work in the attic is adequate lighting, which is not common in attics. So, droplights will be needed and a powerful flashlight should be included.

Don't forget to dress right for attic work—attics can not only be dirty, dusty, and cramped but also hot. And, while doing the sealing and insulating, you will need to kneel on the attic floorboards, so knee pads can prevent pain from crawling and kneeling. Moreover, a lightweight disposable coverall, gloves, and a hat can keep insulation—often irritating and itchy—off your skin.

Another problem with working in attics is exposure to potential safety and health hazards. Because we are in Step 1 of the sealing and insulating process, safety and health should also be a Step 1 one consideration—safety and health first. However, often, as an endless and ongoing number accounts report, individual safety and health considerations are not given a first, second, or third thought until after being exposed to hazards or after being contaminated or injured by some hazard. It's the old "Gee, I never thought about the hazards to my health and safety hazards before I got exposed or injured" statement. Have you ever heard this one?

So, again preplanning and planning in general should always include concerns about individual safety and health. The point is, working in an attic can be hazardous in many ways. For example, consider exposure to excessive heat during summer work in your attic. Not exactly a picnic—actually, eating while working in the attic (or any other location) is not a recommended action. Simply, working in one's attic is not like a hotdog eating contest.

OK, to account for the possibility of heat stress events occurring while working in the attic, it is wise to begin the work early in the day before the attic heats up later as outside temperatures increase. Also, water, the essence of life, the nectar of well-being (we are talking about safe drinking water), is not just recommended but is an essential tool in the attic worker's toolbox. Few would argue about the need for water while working in heated closed environments, but remember, water is great but is useless if you can't breathe. The point? Use a respirator—an OSHA-approved particulate-type respirator or at a minimum, a dust mask to prevent inhalation of hazardous substances. If you are properly equipped with your respirator and have water, then the next precaution is to make sure you are stepping where it is safe—watching your step is called for anytime one is in an attic. You need to walk on joists (a beam used to support floors or roofs) or truss chords, and not on exposed ceiling drywall or insulation. Based on personal experience, you also need to be on the lookout for sharp nails sticking through the roof desk.

Safety and Health Concerns: Subset of Step One

Before we get into the nuts and bolts of insulating and sealing (weatherizing, so to speak) your attic and other areas of your home, and because we are discussing safety and health precautions, it is prudent, in my view, to take our discussion about safety and health precautions to another level. So, let's deviate from the weatherizing, insulating, and sealing of your home and attic so we can point out and discuss important safety and health considerations.

Self-help and professional construction jobs in the weatherization industry (either for new homes and commercial/public buildings or for retrofitting old homes and commercial/public buildings) have increased significantly over the past 15 years. Weatherization practices and jobs include work-related activities ranging from direct installation of weatherization and insulation materials (applications) to assisting applications with installation of weatherization materials to cleanup.

All weatherization applications and materials have some benefits. At the same time, all weatherization applications and materials have some hazards. The hazard information and solutions common to all these applications and materials include the topics of confined spaces, falls, electrical, medical and first aid, ventilation, proper personal protection equipment, heat stress, and respiratory protection.

In insulating and sealing operations, self-help and other workers can be exposed to the hazards identified below:

Fiberglass—has been used for many years in batting as a solid fill insulator. The main concerns for this type of application include skin and eye irritation. In addition, if the fibers are freed from the batting, they can cause respiratory irritation and inflammation. For blow-in application, employers must provide workers with the proper respiratory protection. Inhalation of fiberglass material has been linked to cancer by OSHA, the International Agency for Research on Cancer (IARC), and the National Toxicology Program (NTP).

Cellulose—is the oldest known building insulating material. Dry cellulose can be used in retrofitting old buildings by blowing the cellulose into the wall cavity using bore holes drilled at the top of the walls. Either dry cellulose or wet cellulose applications can be used in new building construction. Cellulose is a respiratory irritant (NTP, 2006). Employers should provide workers with the appropriate dust respirators when using this type of insulation material (NTP, 2006; 29 CFR 1910.1000 Subpart Z; 29 CFR 1926.55 Appendix A). Unless treated with fire-retardants, cellulose can be flammable and should not be used around open flames (Philpot, 1970).

Spray Polyurethane Foam (SPF)[1]—has been used as an insulating material in new construction for many years. However, a new emphasis for retrofitting older buildings to conserve energy has increased the use of SPF at least 60% in the past 5 years. Employers need to ensure that SPF application is carried out in a safe manner to protect workers. SPF contains Isocyanates, which have been reported to be the leading attributable chemical cause of work-related asthma (WRA) (NIOSH, 2004). OSHA has identified the following unique hazards associated with SPF:

- *SPF/Isocyanates*—as mentioned, Isocyanates have been determined to be the leading attributable cause of work-related asthma (NIOSH, 2004). Repeated exposure to Isocyanates has been shown to exacerbate existing asthmatic conditions (Mapp, 2005). Isocyanates are the key materials used to produce polyurethane polymers. These polymers are found in common materials such as polyurethane foams, thermoplastic elastomers, spandex fibers, and polyurethane paints. Jobs that may involve exposure to Isocyanates include painting, foam-blowing, and the manufacture of many polyurethane products, such as chemicals, polyurethane foam, insulation materials, surface coatings, car seats, furniture, foam mattresses, under-carpet padding, packaging materials, shoes, laminated fabrics, polyurethane rubber, and adhesives. Exposure may also occur during the thermal degradation of polyurethane products (e.g., burning or heating at high temperatures).

OSHA has permissible exposure limits (PELs) for Methylene bisphenyl disocyanate (MDI) and 2, 4 toluene diisocyanate TDI of 0.02 ppm. This corresponds to 0.20 mg/m3 for MDI and 0.14 mg/m3for TDI. Health effects of Isocyanate exposure include irritation of skin and mucous membranes, chest tightness, and difficulty breathing. Isocyanates include compounds classified as potential human carcinogens and are known to cause cancer in animals. The main effects of hazardous exposures are sensitization, which can lead to work-related asthma (sometimes called occupational asthma) and other lung problems, as well as irritation of the eyes, nose, throat, and skin.

Below is a list of jobs that may entail potential Isocyanate exposure and materials that may contain Isocyanates. It is important to understand additional sources of Isocyanate exposure, especially for those already sensitized or with asthma, in order to avoid exacerbating an existing asthmatic condition. Because Isocyanate exposure can occur across multiple jobs, it is important to understand the source and routes of exposure.

Potential Job-Related Isocyanate Exposure
- Automotive—paints, glues, insulation, sealants and fiber bonding, truck bed lining
- Casting—foundry cores
- Building and construction—in sealants, glues, insulation materials, fillers
- Electricity and electronics—in cable insulation, PUR-coated circuit boards
- Mechanical engineering—insulation material
- Paints—lacquers
- Plastics—soft and hard plastics, plastic foam, and cellular plastic
- Printing—inks and lacquers
- Timber and furniture—adhesive, lacquers, upholstery stuffing and fabric
- Textile—synthetic textile fibers
- Medical care—PUR casts
- Mining—sealants and insulating materials
- Food industry—packaging materials and lacquers
- *Fires*—there are fire hazards associated with SPF. Employers handling SPF must ensure that electrical equipment is rated for hazardous locations and that there are two ignition sources or open flames in the area. Employers need to make certain that space where SPF is used is well ventilated to avoid accumulation of flammable gases (OSHA 29 CFR 1926.151—General Construction, Fire prevention).

Employers must have the proper fire extinguisher equipment available for use with SPF. For small, local, contained fires, it is acceptable to have extinguishers rated for water, dry extinguishing media, carbon dioxide, or foam.

As with the use of any flammable material, employers must have a plan that outlines the assignments of key personnel in the event of a fire and develop and implement escape procedures for workers. Employers need to make sure this plan is available to workers and train workers on the potential hazards of any worksite.

Polystyrene—is used as a spray-on application like SPF installations. Styrene may be generated during the installation process. Styrene has been shown to cause several health effects when inhaled. These include respiratory irritation and neurological effects. Employers need to provide adequate respiratory protection and protective equipment like SPF installation when using this spray-on application (29 CFR 1910.1000 Subpart Z; 29 CFR 1926.55 Appendix A—Air contaminants; EPA IRIS).

Styrene is also flammable, and the same controls as outlined for SPF should be used (29 CFR 1910.106—Flammable and combustible liquids).

Latex Sealant—is generally used with fiberglass batting in order to provide a better seal. Because latex is a known sensitizer, it can cause allergic skin and respiratory reactions in some individuals. Employers need to provide workers with proper protective equipment and respiratory protection when using this type of product to avoid unnecessary skin, eye, and respiratory exposure.

WEATHER INSULATING/SEALING JOB HAZARDS

Workers in the Weather insulating/sealing industry and self-helpers are exposed to typical workplace hazards, including the following:

HAZARDS AND CONTROLS

- Confined Spaces
- Falls
- Medical and First Aid
- Electrical
- Respiratory Protection
- PPE
- Ventilation

CONFINED SPACE HAZARDS IN WEATHER INSULATING/SEALING ACTIVITIES[2]

As mentioned, self-helpers and professional workers entering confined spaces may be entering into areas that could have accumulated hazardous gases or that could become low oxygen environments, which can pose a risk of asphyxiation. In addition, Isocyanates and other agents contained in SPF are potentially flammable and present a fire hazard when used in spaces that are not properly ventilated. For example, the improper use of SPF in a confined space led to a fire in Falmouth, Massachusetts, (May 2008) when a worker was trapped in an attic space and died. The worker was installing SPF insulation in the attic and did not have proper ventilation or an escape route.

In addition, workers can be exposed to mechanical or electrical equipment inside confined spaces. Some of the areas that workers enter inside a construction/retrofitting site may be considered a confined space.

Employers need to look at the spaces that workers enter to determine whether they meet OSHA's definition of confined space, and, if so, whether the space has any other associated hazards that make it a "permit-required" confined space. If workers need to enter a "permit-required" confined space, then the employer must develop entry procedures, including an entry permit, and train the workers. For further information on confined space hazards and appropriate safe work practices, they should be consulted along with pertinent OSHA standards.

FALL HAZARDS IN WEATHER INSULATING/SEALING JOBS[3]

Workers applying blow-in or spray-on materials at elevated locations may be exposed to fall hazards. Workers applying blow-in or spray-on materials use specialized equipment that includes long hoses and electrical power lines. These can create trip and fall hazards on stairs, platforms, and scaffolding, as well as flat surfaces. Employers must protect their workers from fall hazards (training—29 CFR 1926.1060; Protection—scaffolds 29 CFR 1926.454 and duty to have fall protection—29 CFR 1926.501).

MEDICAL AND FIRST AID IN WEATHER INSULATING/SEALING JOBS[4]

Weatherization and insulation materials can be used at new construction sites as well as remodeling sites. New home construction sites can sometimes be found in more remote places without easy access to medical or clinical personnel. If this is the case, the employers need to ensure that medical personnel are available for advice and consultation, and that someone who is trained is available to provide first aid.

ELECTRICAL SAFETY IN WEATHER INSULATING/SEALING JOBS[5]

Workers in the weatherization industry may be exposed to potential electrical hazards present in their work environment, which makes them more vulnerable to the danger of electrocution and to arc flash hazards. Equipment used for blow-in/spray-on applications may require the use of long electrical cords or extension cords that can be subject to wear. It is important to inspect all electrical lines and maintain them in good working order.

Specifically because some materials such as styrene and components of SPF are flammable, it is essential that employers ensure that electrical equipment provided to workers is safe for use in hazardous (classified) locations, and it must be maintained properly to prevent sparking or ignition of the flammable vapors when installing this material. All electrical wiring, including extension cords, needs to be maintained properly to ensure safe working conditions. Employers must comply with OSHA's electrical standards to protect workers from these hazards.

RESPIRATORY HAZARDS IN WEATHER INSULTING/SEALING JOBS

Many blow-in/spray-on applications contain chemicals or materials that can cause irritation or sensitization. These have been highlighted in the unique hazards section for each application on the main weatherization web page. However, some applications have similar respiratory protection requirements. For blow-in materials such as fiberglass and cellulose, if engineering controls are not feasible, employers are required to supply workers with appropriate dust respirators if exposures exceed the PEL (29 CFR 1910.1000 Subpart Z, 29 CFR 1926.55 Appendix A).

SPF Respiratory Protection Recommendations—workers in the SPF industry can be exposed to inhalation hazards from Isocyanates and other hazardous chemical vapors or dusts. OSHA requires a hierarchy of controls, under which employers must first implement engineering controls (including elimination, substitution) and/or administrative controls whenever possible. If such controls are not feasible to achieve full compliance, personal protective equipment or any other protective measures must be used to ensure the employees are not exposed to air contaminants above permissible limits. However, for some situations engineering controls are not adequate. For the spray foam applicator and the helper (who is standing in proximity and/or may switch duties with spray applicator), a respirator with full face mask should be provided and worn during application and cleanup to avoid skin contact. It is recommended that employers provide a supplied-air respirator for both the safety and comfort of the worker since heat stress can become an issue.

If using a cartridge filter respirator, a full-face respirator with an organic vapor sorbent and particulate filter must be used. OSHA requires that employers develop a changeout schedule for the respirator cartridges to avoid saturating the filter capacity. (See 29 CFR 1910.134 and 29 CFR 1926.103.)

For non-applicators, respirators should be used to supplement worker protection only when engineering controls (for example, ventilation) have been shown to be inadequate or during the interim period when engineering controls are being installed.

PERSONAL PROTECTIVE EQUIPMENT FOR WEATHER INSULATING/SEALING JOBS[6]

Workers in the SPF industry can be exposed to inhalation hazards from Isocyanates and other hazardous chemical vapors or dusts. OSHA requires a hierarchy of controls, under which employers must first implement engineering controls (including elimination, substitution) and/or administrative controls whenever possible. If such controls are not feasible to achieve full compliance, personal protective equipment or any other protective measures must be used to ensure the employees are not exposed to air contaminants above permissible limits. For the spray foam applicator and the helper (who is standing in proximity and/or may switch duties with spray applicator), a respirator with full face mask should be provided and worn during application and clean-up. In addition, full saran-coated suits/coveralls with appropriate gloves and shoe covers should be provided and worn during application and cleanup. It is highly recommended that no skin exposure occur during these work processes (Bello, 2007).

It is recommended that workers in trimming or sanding operations should be provided with and wear appropriate respirators, long sleeves, and gloves. Goggles or other eye protective equipment is required.

Protective Equipment for SPF Application

- Hand protection—chemical-resistant protective gloves, chloroprene rubber (Neoprene), nitrile rubber (Buna N), chlorinated polyethylene, polyvinylchloride (Pylox), butyl rubber, fluoroelastomer (Viton).
- Eye protection—tightly fitting safety goggles (chemical goggles). Wear face shield if splashing hazard exists. Full face mask for applicator.
- Body protection—saran-coated material (full body suit with hood for applicator).

General Safety and Hygiene Measures

- Wear protective clothing as necessary to prevent contact. Eye wash fountains and safety showers must be easily accessible. Observe the appropriate PEL value. Wash soiled clothing immediately. Contaminated equipment or clothing should be disposed of or cleaned thoroughly after each use. Use protective equipment while cleaning to avoid contamination.
- Respirators should be used by the workers for protection only when engineering controls (for example, ventilation) have been shown not to be feasible or during the interim period when engineering controls are being installed.

Ventilation in Weather Insulating/Sealing Jobs[7]

OSHA requires engineering controls to be used whenever feasible to reduce exposure to Isocyanates (29 CFR1910.1000 and 29 CFR 1926.55—Air contaminants), which is the major component of SPF. Areas where SPF is being sprayed should be separated from other portions of the building. In addition, ductwork should be sealed to prevent the spread of fumes or vapors to other areas. Air handlers (fans) should move air about the room in one direction and move vapors/air contaminants away from operators and other workers in the area. Air should be vented to the outside using filters like those used for truck bed-liner booths (use a similar system as used in truck bed-liner industry). OSHA requires that all workers be informed of the presence of Isocyanates in SPF and be provided with proper training.

History: Fatalities and Incidents

OSHA has identified several fatalities and incidents due to severe asthmatic attacks and fire/explosions associated with the use of Isocyanate-containing materials.

- A 41-year-old Springfield, Massachusetts, worker was killed when the spray foam chemicals he was spraying in a home attic caught fire. A worker from a Vermont insulation company ran a hose from two 50-gallon drums of chemicals outside the house into the attic where he was applying the insulator. It is believed that the vapors of the spray chemicals then ignited and engulfed the attic. After smelling smoke, two coworkers rushed upstairs to remove the worker but couldn't, because the flames and smoke were too intense. Firefighters were unable to reach the man by placing a ladder on the porch roof. After breaking into the side window to the attic, firefighters pulled the man's unconscious body out of the house and performed CPR. The efforts were unsuccessful. The worker was pronounced dead at the Springfield hospital that night (Gouveia, 2008).
- A maintenance worker repairing a foaming system at a polyurethane foam manufacturing plant developed respiratory symptoms associated with Isocyanates exposure. Detectable MDI concentrations were discovered in the workplace. There was no effective ventilation and dermal protection when investigations revealed aerosols and vapors near the faces of the workers. The worker quit his job after being diagnosed with Isocyanate-induced hypersensitivity pneumonitis. Years after leaving the plant, he continued to experience symptoms including cough, weakness, sweat, muscle aches, shortness of breath, and loss of lung function. The illness worsened over time, and eventually led to his death (NIOSH, 1996).
- A 45-year-old worker had a fatal asthma attack after spraying an MDI-based bed liner onto a vehicle interior. The worker was wearing a half-mask, a supplied-air respirator, latex gloves, and coveralls. The room had two curtains pulled together to contain the spray, a fan at the door to provide air circulation, and no local exhaust ventilation. After disconnecting the respirator and leaving the room, the worker began developing acute respiratory symptoms. He was taken to the hospital where he then went into cardiac arrest. The county medical examiner stated that the worker had died of an "acute asthmatic reaction due to inhalation of chemicals." The health and work history of the patient suggests that MDI sensitization played a significant part in the fatality (NIOSH, 2006).
- Public school officials at a large metropolitan school district became suspicious after several staff members developed asthma symptoms. NIOSH investigators were called in to inspect the school and discovered several recent Isocyanate foam and coating material applications. The staff reported odors from the materials when they were being applied. Air sampling tests indicated release of Isocyanates and potential for exposure (NIOSH, 2006).

SEALING ATTIC AIR LEAKS

When I went to seal my attic air leaks, I first composed a materials checklist based on guidance provided in the USEPA's Energy Star Program. The checklist for the materials I carried into my attic and the tools in my toolbox includes:

- Batt or roll of unfaced fiberglass insulation and large garbing bags (for stuffing open stud cavities behind knee walls and in dropped soffits), By the way, knee walls are those short walls in a room with a sloped ceiling. They are usually formed when the room ceiling follows the roofline of a house.
- Roll of reflective foil insulation or other blocking material such as drywall or pieces of rigid forma insulation to cover soffits, open walls, and larger holes.
- For sealing holes one-quarter-inch or less silicone or acrylic latex caulk and caulk gun.
- Several cans of expanding spray form insulation for filling large gaps in the range of one-quarter inch to 3 inches.
- Special high-temperature (heat-resistant) caulk to seal around flues and chimneys.
- Roll of 14-inch aluminum flashing to keep insulation away from the flue pipe.
- Sheet metal scissors and retractable utility knife.
- Hammer and nails for holding covering materials in place (or a staple gun) and a tape measure.
- Safety glasses, gloves, and dust mask.
- Portable safety light or flashlight.
- Extra boards to walk on, if needed.
- Large bag or bucket to carry and haul materials.

STEP 2

After preparing all the insulating and sealing materials, hauling them into my attic, and after paying attention to recommended safety and health precautions, I was prepared to take Step 2 in insulating and sealing my attic.

So, following the guidance in the USEPA's Energy Star Program, I proceeded to plug the big holes first. This step makes sense because finding and sealing all the little holes in my attic is secondary to my goal of energy savings by plugging the large holes first. When I entered the attic, the first thing I did was to refer to my rough drawing of the floor plan. Using my drawing, I located those areas where leakage was likely to be the greatest, that is, where walls—inner and outer—meet the attic floor, dropped soffits (dropped ceiling areas), and behind and under attic knee walls. I looked for dirty insulation—this indicates that air is moving through it. The dropped soffits were filled with insulation and hard to see. I pushed back the insulation and scooped it out of the soffits. I knew I had to place the insulation back over the soffits once I plugged the stud cavities. Then I plugged and covered the soffits. I had a few "can" lights and my open soffits.

Recessed "can" lights (aka high-hats or recessed downlights) are a big source of air leaks, and there is no easy solution. Recessed "can" lights are metal light fixtures (or cans) that are inset into the ceiling. These fixtures were a big source of leaks because they were installed on the upper floor of my house. My "can" lights look terrific, but they protrude (all 6 of them) into my attic space, making my home less energy efficient. These recessed lights in the ceiling of my second-story house created open holes into my attic that allowed unwanted heat flow between conditioned and unconditioned spaces. In the summer, hot attic air was making the rooms warmer, and in the winter "can" lights draw warm air up into my attic. Both the warm air leakage and the heat from the lights caused problems in cold times, the heat melted snow on the roof and forced ice dams (the water refroze at the roof edge). I also had a couple of recessed "can" lights in two upper-story bathrooms and the warm, moist air was leaking into the attic.

One thing about my "can" lights that I was warned about by Energy Star was that I needed to keep all my attic insulation at least 3 inches from the light. I looked at this issue and decided to use a piece of circular metal flashing around the light as a dam to keep the insulation away from the light.

Regarding plugging the large holes in my attic, what I did was cut a 16-inch-long piece from a batt of unfaced fiberglass insulation and fold it into the bottom of a 13-gallon plastic garbage bag. I folded the bag and stuffed it into the open stud cavity. I had to add more insulation to the bag because it did not fit tightly.

In covering my dropped soffits, I first removed insulation from the soffit. Then I cut a length of reflective foil blocking material (rigid foam board worked well for me) a few inches longer than the opening to be covered. I applied a bead of caulk around the opening. I then sealed the foil to the frame with the caulk and stapled it in place.

At the knee walls, I cut a 24-inch-long piece from a batt of fiberglass insulation and placed it at the bottom of a 13-gallon plastic garbage bag. I folded the bag over and stuffed it into the open joist spaces under the wall. Then I covered the insulation.

DID YOU KNOW?

Some attics have lightweight, pea-size vermiculite insulation, which may contain asbestos. Thus, the insulation containing vermiculite insulation must be left alone unless you have had a lab test the insulation to ensure its safety. Otherwise, and if necessary, professional licensed asbestos personnel should be called to handle the hazard correctly.

While insulating and sealing my attic, I also found that the opening around my furnace chimney and water heater flue were a main source of warm air moving in my attic. Because the pipe gets hot, building codes usually require 1 inch of clearance from metal flues (2 inches from masonry chimneys—mine is metal pipe) to any combustible material, including insulation. I installed metal flashing around the openings using high-temperature (heat-resistant) caulk. Also, before I pushed insulation back into place, I built a metal dam to keep it away from the pipe. Note that if my chimney were masonry, I would have performed the same technique. For my round metal pipe flue, I cut half-circles out of two pieces of aluminum flashing so they overlapped about 3 inches in the middle. I pressed the flashing metal into a bead of high-temperature material and stapled it into place. I had wood to staple to, but if instead I had had drywall, I would have nailed it directly into the drywall; however, I would make sure I did not nail directly through the drywall and did not staple through the drywall.

Caution: A cautionary note is called for here: some self-helpers use foam instead of high-temperature caulk. Do not use spray foam. For a furnace/water heater made of balkanized metal, seal with aluminum flashing with high-temperature caulk. If you have a chimney flue/vent/pipe made of masonry/metal, use aluminum flashing and high-temperature silicone caulk. If the piping in the attic is made from cast iron or PVC, expanding foam or caulk can be used, depending on the size of the gap.

After I plugged the large holes in my attic, I had to put on my Sherlock Holmes hat to investigate, to find, and to seal the small gaps. Not easy. It took me a while to find what I could find. The problem was almost all the small leaks were under the insulation. Again, I did find what I could not see, but I am not sure I got them all. Anyway, I found darkened insulation that led me to the ones I found. This darkened insulation resulted from filtering dirty air in my home. During cold weather, I discovered frosty areas in the insulation caused by warm moisture that had condensed and then froze as it hit the cold attic air. I also found water staining in these same areas. I determined that even though the insulation was dirty, it was still OK to use. So, I left it be. After sealing the areas with expanding foam and caulk, I pushed the insulation back into place. Some of my insulation is blown insulation, so I used a small rake to level it back into place.

I did find gaps in electrical cables and connection boxes and a couple of vent pipes in the attic that I filled with expanding foam (I could have used caulk, but my preference is expanding foam where applicable). After I read the instructions on the expanding foam can, I had to be careful in applying the expanding foam, so I used gloves and made sure none got on my clothing—the foam is quite sticky and almost impossible to remove once it sets. So, I waited for the applied foam to dry and then I covered the areas with insulation. I also had one 3-inch-wide plumbing pipe, so I stuffed the space around it with fiberglass insulation as a backer for expanding foam.

While sealing and insulating my heating and cooling ducts in my attic, I checked the duct connects for leaks by turning on my heating and cooling system fan and feeling for leaks—I only found a few leaks that I sealed the joints with foil tape. I could have used mastic, which is a good duct sealer. It is a paste and is more durable than foil tape, but I found it easier, in my case, to apply foil tape. Do not seal your ducts with duct tape—grey duct tape is not to be used because it fails quickly. While checking for duct leaks, I decided to insulate the ducts. I used zip ties to secure them in place. I sealed the ducts first and then I used R-6 insulation material where needed.

OK, let's review the materials I used to insulate and seal leaks and ducts in my attic:

- Sealant using metal-backed foil tape
- Duct insulation material rated at R-6
- Zip ties to hold duct insulation in place
- Gloves, safety glasses, mask, flashlight

DID YOU KNOW?

A point of caution: you should check to make sure that there is no carbon monoxide. The point is that, after you make your insulation and sealing green energy improvements that result in a tighter house, there can be an increased opportunity for carbon dioxide (CO) to build up if your gas-burning appliances are not venting properly. The prudent thing to do is to have your heating and cooling technician check your combustion appliances, such as your gas or oil-fired furnace, water heater, and dryer for proper venting.

After finishing my work in my attic, I sealed the access hatch with self-sticking weather stripping.

STEP 3–IN THE BASEMENT

When I finished plugging holes in my attic, I was not done with properly weatherizing by insulating and sealing my attic. I also have a basement and there is a need to seal basement air leaks. The truth be told, I found that outside air drawn in through my basement leaks was being exacerbated by the chimney effect created by the leaks I had in my attic. As hot air generated by the basement furnace rose through the house and into the attic through leaks, cold outside air gets drawn in through basement leaks to replace the displaced air. This was making my home feel drafty and contributing to higher energy bills—not a good green energy practice. Therefore, after sealing my attic air leaks, I had to complete the job by sealing basement leaks, to stop the chimney effect.

Well, needless to say I had to determine where the basement leaks were coming from. After checking the USEPA Energy Star program, I found that a common area of air leakage in the basement is along the top of the basement wall where cement or block comes in contact with the wood frame. After following the Energy Star advice, I had my Eureka! moment when I celebrated finding the source of my air leaks. These leaks were easily fixed in the portions where my basement was not finished. I found that since the top of the wall is above ground, outside air was being drawn in through cracks and gaps where the house framing sits on top of the foundation. The perimeter framing is called the rim (or band) joist. In my basement, the above-floor joists end at the rim joist, creating multiple cavities along the length of the wall, and many opportunities for leakage (which turned out to be the case in my basement).

I knew I could not see cracks in the rim and joints, so it was best to seal up the top and bottom of the inside of the cavity. Also, I sealed the rim joist at points where my bump out bay windows were hanging off the foundation. These areas provide greater opportunities for air leakage and heat loss. I used caulk to seal all the one-quarter-inch cracks and less. Those cracks that were more than one-quarter inch, up to about 3 inches, I filled with spray foam. The basement ceiling to the floor above was my next sealing target. In particular, I sought out holes for wires, water supply pipes, water drainpipes, the plumbing vent stack (for venting sewer gases), and finally the furnace flue (for venting furnace exhaust).

WEATHER INSULATING/SEALING JOB HAZARDS

Jobs in the weatherization industry (either for new homes and commercial/public buildings or for retrofitting old homes and commercial/public buildings) have increased significantly over the past 20 years. Weatherization jobs include work-related activities ranging from direct installation of weatherization and insulation materials (applications) as detailed above, to assisting applications with installation of weatherization materials to cleanup. There is little doubt about the benefits of weatherization of your residence or other structure—a bona fide green project. However, it is important to keep in mind that there are potential hazards involved with weatherizing activities and a few of them are described, beginning with Isocyanates.

OSHA has permissible exposure limits (PELs) for Methylene biphenyl disocyanate (MDI) and 2, 4 toluene diisocyanate TDI of 0.02 ppm. This corresponds to 0.20 mg/m3 for MDI and 0.14 mg/m3for TDI. Health effects of Isocyanate exposure include irritation of skin and mucous membranes, chest tightness, and difficulty breathing. Isocyanates include compounds classified as potential human carcinogens and are known to cause cancer in animals. The main effects of hazardous exposures are sensitizations which can lead to work-related asthma (sometimes called occupational asthma) and other lung problems, as well as irritation of the eyes, nose, throat, and skin.

Below is a list of jobs with potential Isocyanate exposures and materials that may contain Isocyanates. It is important to understand additional sources of Isocyanate exposure, especially for those already sensitized or with asthma, in order to avoid exacerbating an existing asthmatic condition. Because Isocyanate exposure can occur across multiple jobs, it is important to understand the source and routes of exposure.

Potential Jobs-Related Isocyanate Exposures
- Automotive—paints, glues, insulation, sealants and fiber bonding, truck bed lining
- Casting—foundry cores
- Building and construction—in sealants, glues, insulation materials, fillers
- Electricity and electronics—in cable insulation, Polyurethane (aka PUR)-coated circuit boards
- Mechanical engineering—insulation material
- Paints–lacquers
- Plastics—soft and hard plastics, plastic foam, and cellular plastic
- Printing—inks and lacquers
- Timber and furniture—adhesive, lacquers, upholstery stuffing and fabric
- Textile—synthetic textile fibers
- Medical care—PUR casts
- Mining—sealants and insulating materials

- Food industry—packaging materials and lacquers
- *Fires*—there are fire hazards associated with spray polyurethane foam (SPF). Employers handling SPF must ensure that electrical equipment is rated for hazardous locations and that there are two ignition sources or open flames in the area. Employers need to make certain that space where SPF is used is well ventilated to avoid accumulation of flammable gases. (OSHA 29 CFR 1926.151—General Construction, Fire prevention)

    Employers must have the proper fire extinguisher equipment available for use with SPF. For small, local, contained fires it is acceptable to have extinguishers rated for water, dry extinguishing media, carbon dioxide, or foam.

    As with the use of any flammable material, employers must have a plan that outlines the assignments of key personnel in the event of a fire and develop and implement escape procedures for workers. Employers need to make sure this plan is available to workers and train workers on the potential hazards of any worksite.

Polystyrene—is used as a spray-on application similar to SPF installations. Styrene may be generated during the installation process. Styrene has been shown to cause several health effects when inhaled. These include respiratory irritation, and neurological effects. Employers need to provide adequate respiratory protection and protective equipment similar to SPF installation when using this spray-on application (29 CFR 1910.1000 Subpart Z; 29 CFR 1926.55 Appendix A—Air contaminants; EPA IRIS).

Styrene is also flammable and the same controls as outlined for SPF should be used (29 CFR 1910.106—Flammable and combustible liquids).

Latex Sealant—is generally used with fiberglass batting in order to provide a better seal. Because latex is a known sensitizer it can cause allergic skin and respiratory reactions in some individuals. Employers need to provide workers with proper protective equipment and respiratory protection when using this type of product to avoid unnecessary skin, eye, and respiratory exposure.

THE BOTTOM LINE

Weatherizing by insulating and sealing homes and workplaces is a huge green energy practice simply because it reduces the need for energy and also the polluting effects of using more energy. Workers in the weather insulating/sealing industry are exposed to typical workplace hazards, including the following:

HAZARDS AND CONTROLS

- Confined Spaces
- Falls
- Medical and First Aid
- Electrical
- Respiratory Protection

- PPE
- Ventilation

Confined Space Hazards in Weather Insulating/Sealing Jobs[8]

As mentioned, workers entering confined spaces may be entering into areas that could have accumulated hazardous gases or that can become low oxygen environments, which can pose a risk of asphyxiation. In addition, Isocyanates and other agents contained in the SPF are potentially flammable and present a fire hazard when used in spaces that are not properly ventilated. For example, the improper use of SPF in a confined space led to a fire in Falmouth, Massachusetts, (May 2008) when a worker was trapped in an attic space and died. The worker was installing SPF insulation in the attic and did not have proper ventilation or an escape route.

In addition, workers can be exposed to mechanical or electrical equipment inside confined spaces. Some of the areas that workers enter inside a construction/retrofitting site may be considered a confined space.

Employers need to look at the spaces that workers enter to determine if they meet OSHA's definition of confined space, and, if so, whether the space has any other associated hazards that make it a "permit-required" confined space. If workers need to enter a "Permit-required" confined space, then the employer must develop entry procedures, including an entry permit, and training the workers. For further information on confined space hazards and appropriate safe work practices discussed in this text consult pertinent OSHA standards.

Fall Hazards in Weather Insulating/Sealing Jobs[9]

Workers applying blow-in or spray-on materials at elevated locations may be exposed to fall hazards. Workers applying blow-in or spray-on materials use specialized equipment which includes long hoses and electrical power lines. These can create trip and fall hazards on stairs, platforms and scaffolding, as well as flat surfaces. Employers must protect their workers from fall hazards (training–29 CFR 1926.1060; Protection—scaffolds 29 CFR 1926.454 and duty to have fall protection—29 CFR 1926.501).

Medical and First Aid in Weather Insulating/Sealing Jobs[10]

Weatherization and insulation materials can be used at new construction sites as well as remodeling sites. New home construction sites can sometimes be found in more remote places without easy access to medical or clinical personnel. If this is the case, the employers need to ensure that medical personnel are available for advice and consultation, and that someone who is trained is available to provide first aid. See OSHA's Medical and First Aid Safety and Health Topics Page and Tab E of this text for standards and safety practices.

Electrical Safety in Weather Insulating/Sealing Jobs[11]

Workers in the weatherization industry may be exposed to potential electrical hazards present in their work environment, which makes them more vulnerable to the danger of electrocution, and arc flash hazards. Equipment used for blow-on/spray-on applications may require the use of long electrical cords or extension cords that can be subject to wear. It is important to inspect all electrical lines and maintain them in good working order.

Specifically because some materials such as styrene and components of SPF are flammable, it is essential that employers ensure that electrical equipment provided to workers is safe for use in hazardous (classified) locations and must be maintained properly to prevent sparking or ignition of the flammable vapors when installing this material. All electrical wiring, including extension cords, needs to be maintained properly to ensure safe working conditions. Employers must comply with OSHA's electrical standards to protect workers from these hazards.

Respiratory Hazards in Weather Insulting/Sealing Jobs

Many blow-on/spray-on applications contain chemicals or materials which can cause irritation or sensitization. These have been highlighted in the unique hazards section for each application on the main weatherization web page. However, some applications have similar respiratory protection requirements. For the blow-in materials such as fiberglass and cellulose, if engineering controls are not feasible, employers are required to supply workers with appropriate dust respirators if exposures exceed the PEL (29 CFR 1910.1000 Subpart Z, 29 CFR 1926.55 Appendix A).

SPF Respiratory Protection Recommendations—workers in the SPF industry can be exposed to inhalation hazards from Isocyanates and other hazardous chemical vapors or dusts. OSHA requires a hierarchy of controls, under which employers must first implement engineering controls (including elimination, substitution) and/or administrative controls whenever possible. If such controls are not feasible to achieve full compliance, personal protective equipment or any other protective measures must be used to ensure the employees are not exposed to air contaminants above permissible limits. However, for some situations engineering controls are not adequate. For the spray foam applicator and the helper (who is standing in close proximity and/or may switch duties with spray applicator) a respirator with full face mask should be provided and worn during application and clean-up to avoid skin contact. It is recommended that employers provide a supplied-air respirator for both safety and comfort of the worker since heat stress can become an issue.

If using a cartridge filter respirator, a full-face respirator with an organic vapor sorbent and particulate filter must be used. OSHA requires that employers develop a changeout schedule for the respirator cartridges to avoid saturating the filter capacity. See 29 CFR 1910.134 and 29 CFR 1926.103.

For non-applicators, respirators should be used to supplement worker protection only when engineering controls (for example, ventilation) have been shown to be inadequate or during the interim period when engineering controls are being installed.

Personal Protective Equipment for Weather Insulating/Sealing Jobs[12]

Workers in the SPF industry can be exposed to inhalation hazards from Isocyanates and other hazardous chemical vapors or dusts. OSHA requires a hierarchy of controls, under which employers must first implement engineering controls (including elimination, substitution) and/or administrative controls whenever possible. If such controls are not feasible to achieve full compliance, personal protective equipment or any other protective measures must be used to ensure the employees are not exposed to air contaminants about permissible limits. For the spray foam applicator and the helper (who is standing in close proximity and/or may switch duties with spray applicator) a respirator with full face mask should be provided and worn during application and clean-up. In addition, full saran-coated suits/coveralls with appropriate gloves and shoe-covers should be provided and worn during application and clean-up. It is highly recommended that no skin exposure occurs during these work processes (Bello, 2007).

It is recommended that workers in trimming or sanding operations should be provided and wear appropriate respirators, long sleeves and gloves. Goggles or other eye protective equipment is required.

Protective Equipment for SPF Application

- Hand protection—chemical-resistant protective gloves, chloroprene rubber (Neoprene), nitrile rubber (Buna N), chlorinated polyethylene, polyvinylchloride (Pylox), butyl rubber, fluoroelastomer (Viton).
- Eye protection—tightly fitting safety goggles (chemical goggles). Wear face shield if splashing hazard exists. Full face mask for applicator.
- Body protection—saran-coated material (full body suit with hood for applicator).

General Safety and Hygiene Measures

- Wear Protective clothing as necessary to prevent contact. Eye wash fountains and safety showers must be easily accessible. Observe the appropriate PEL value. Wash soiled clothing immediately. Contaminated equipment or clothing should be disposed of or cleaned thoroughly after each use. Use protective equipment while cleaning to avoid contamination.
- Respirators should be used by the workers for protection only when engineering controls (for example, ventilation) have been shown not to be feasible or during the interim period when engineering controls are being installed.

Ventilation in Weather Insulating/Sealing Jobs[13]

OSHA requires engineering controls to be used whenever feasible to reduce exposures to Isocyanates (29 CFR1910.1000 and 29 CFR 1926.55—Air contaminants), which is the major component of SPF. Areas where SPF is being sprayed should be separated from other portions of the building. In addition, duct work should be sealed to prevent

spread of fumes or vapors to other areas. Air handlers (fans) should move air about
the room in one direction and move vapors/air contaminants away from operators and
other workers in the area. Air should be vented to the outside using filters similar to
those used for truck bed-liner booths (use similar system as used in truck bed-liner
industry). OSHA requires that all workers be informed of the presence of Isocyanates
in SPF and be provided with proper training.

NOTES

1 Based on material contained in OSHA's *Green Job Hazards: Weather Insulating/Sealing: Chemical Hazards—SPF/Isocyanates.* Accessed 7/18/23 @ www.osha.gov/dep/greenjobs/spf_chemical.html
2 Based on information from OSHA's *Green Job Hazards: Weather Insulating/Sealing—Confined Spaces.* Accessed 7/18/23 @ www.osha.gov/dep/greenjobs/weather_confined.html
3 Based on information from OSHA's *Green Job Hazards: Weather Insulating/Sealing—Falls.* Accessed 7/18/23 @ www.osha.gov/dep/greenjobs/weather_falls.html
4 Based on information from OSHA's *Green Job Hazards: Weather Insulating/Sealing—Medical and First Aid.* Accessed 7/19/23 @ www.osha.gov/dep/greenjobs/weather_medical.html
5 Based on information from OSHA's *Green Job Hazards: Weather Insulating/Sealing—Electrical.* Accessed 7/19/23 @ www.osha.gov/dep/greenjobs/weatehr_electrical.html
6 Based on information from OSHA's *Green Job Hazards: Weather Insulating/Sealing—Personal Protective Equipment.* Accessed 7/19/23 @ www.osha.gov/dep/greenjobs/weather_ppe.html
7 Based on information from OSHA's *Green Job Hazards: Weather Insulating/Sealing--Ventilation.* Accessed 7/20/23 @ www.osha.gov/dep/greenjobs/weather_ventilation.html
8 Based on information from OSHA's *Green Job Hazards: Weather Insulating/Sealing—Confined Spaces.* Accessed 11/18/11 @ www.osha.gov/dep/greenjobs/weather_confined.html
9 Based on information from OSHA's *Green Job Hazards: Weather Insulating/Sealing—Falls.* Accessed 11/18/11 @ www.osha.gov/dep/greenjobs/weather_falls.html
10 Based on information from OSHA's *Green Job Hazards: Weather Insulating/Sealing—Medical and First Aid.* Accessed 11/19/11 @ www.osha.gov/dep/greenjobs/weather_medical.html
11 Based on information from OSHA's *Green Job Hazards: Weather Insulating/Sealing—Electrical.* Accessed 11/19/11 @ www.osha.gov/dep/greenjobs/weatehr_electrical.html
12 Based on information from OSHA's *Green Job Hazards: Weather Insulating/Sealing—Personal Protective Equipment.* Accessed 11/19/11 @ www.osha.gov/dep/greenjobs/weather_ppe.html
13 Based on information from OSHA's *Green Job Hazards: Weather Insulating/Sealing—Ventilation.* Accessed 11/20/21 @ www.osha.gov/dep/greenjobs/weather_ventilation.html

REFERENCES AND RECOMMENDED READING

Bello, D, Herrick, C.A., Smith T.J., Woskie, S.R., Streicher, R.P., Cullen, M.R., Liu, Y., and Redich, C.A. 2007. Skin exposure in Isocyanates: Reasons for concern. *Evironmenatl Health Perspectives* Mar.; **115**(3): 328–335.
Gouveia, A. 2008. "Green" insulation suspected in fatal fire. *Cape Cod Times*, May 20.
Mapp, C.E., Boschetto, P., Maestrelli, P., and Fabbri, L.M. 2005. Occupational Asthma. *American Journal of Respiratory and Critical Care Medicine* **172**; 28/0–305.
Morgan, D.L. 2006. NTP Toxicity Study Report on the atmospheric characterization, particle size, chemical composition, and workplace exposure assessment of cellulose insulation. *Toxicity Report Series* Aug.; **74**: 1–62, A1–C2.
National Institute for Occupational Safety and Health (NIOSH). 1996. *Preventing Asthma and Death from Diisocyanate Exposures.* pub. No. 96-111. Atlanta, Georgia: Centers for Disease Control.

National Institute for Occupational Safety and Health (NIOSH). 2004. *Worker Health Chartbook 2004*. NIOSH Publication Number 2004-146.

Philpot, C.W. 1970. Influence of mineral content on the pyrolysis of plant materials. *Forest Science* **16**: 461–471.

US Environmental Protection Agency (USEPA). 2009. *Why Seal and Insulate*. Accessed 6/6/23 @ www.-energystar-gov/safeathome/seal-insulate/why-seal-and-insulation

12 Recycling

WASTE MANAGEMENT AND RECYCLING JOB HAZARDS

Based on personal research and observation, I have found that waste, and how we manage it, typically ends up toward the bottom of the environmental policy agenda. When the truth is being told it hurts, but it is absolutely necessary if we are to maintain the lifestyles that many of us in free nations enjoy today. So, let's point out some "truth being told" about waste management. I do this by pointing out and discussing the information provided below—you might think to yourself: "What the hell does any of this recycling information and discussion have to do with green energy practices?" Well, hang on and follow the discussion below—then think about it.

Think!

In 2008, the US Environmental Protection Agency (USEPA) estimated that of the 250 million tons of waste generated in the United States, approximately one-third, or 83 million tons, was recycled or composted. Since 1985, the percentage of waste recycled in the United States has doubled, and the trend is likely to continue. As the recycling industry continues to grow, so do the number of available jobs, each with its own safety and health risks.

RECYCLING: WHAT IT IS ALL ABOUT

Recycling and energy conservation and green energy are all about, simply, conserving energy—energy of any type and from any source. Recycling is the process of collecting and processing materials that would otherwise be thrown away as trash and turning them new products. Turning waste into new products is all about reuse. Reuse is all about conserving materials and, more importantly, about conserving energy. Recycling and reuse often save energy and natural resources—basic green energy practice. The natural resources include land, plants, minerals, and water. When we use materials more than once, we conserve natural resources and energy—definitely a green energy practice. Consider the following:

- When we do not have to increase our mining activities to resupply materials needed for new appliances and other applications, it is the environment that wins.

DOI: 10.1201/9781003439059-15

- Making a product from recycled materials almost always requires less energy and it is the environment that wins. For instance, think about aluminum. That is, think about recycling aluminum cans to make new aluminum cans, which results in using 95% less energy than using bauxite ore, the mineral substance that raw aluminum is made from; when this occurs the environment wins.
- According to the USEPA, recycling one ton of paper would (USEPA, 2023)
  - Save enough energy to power the average American home for six months
  - Save 7,000 gallons of water
  - Save 3.3 cubic yards of landfill space
  - Reduce greenhouse gas emissions by one metric ton (2,205 pounds) of carbon equivalent.
  - USEPA (2023) provides another example of energy savings by recycling:
  - Recycling one ton of office paper can save the energy equivalent of consuming 322 gallons of gasoline.
  - Recycling just one ton of aluminum cans conserves more than 152 million Btu, the equivalent of 1,024 gallons of gasoline or 21 barrels of oil consumed.
  - Plastic bottles are the most recycled plastic product in the United States as of 2018.

Think about this one too:

Recycling just 10 plastic bottles saves enough energy to power a laptop for more than 25 hours.

RECYCLING: THE CHALLENGES

There is very little doubt about the benefits of recycling and, as a result, the creation of more jobs to enhance the environment. But there challenges including the following (USEPA, 2023):

- Most Americans want to recycle, as they, along with this author, believe recycling provides an opportunity for them to be responsible caretakes of the Earth. Let's face it. Most rational people do not want to pollute, contaminate, spoil, and defile their own backyards, so to speak. However, the fly in the ointment is that it can be difficult for consumers to understand what materials can be recycled, how materials can be recycled, and where to recycle different materials.
- Another problem, or challenge, is that America's recycling infrastructure has not kept pace with today's waste stream. Communication between the manufacturers of new materials and products and the recycling industry needs to be enhanced to prepare for the optimal recycling of new materials.
- Domestic markets for recycled materials need to be strengthened. In the past, some of the recycled material generated in the United States have been exported internationally. It has been the changing of international policies that has not limited the spource of materials. What is needed is better integration of recycled materials and end-of-life management into product and packaging designs. Also, there is the need to improve communication among the different

sectors of the recycling system to strengthen existing materials markets and to develop new innovation markets.

- Metrics are important in that entities across the recycling system agree that more consistent measurement methodologies are needed to measure recycling system performance.
- Another challenge with recycling is to keep workers in the industry safe from the hazards involved with recycling. The next section addresses this issue by providing real-world incident accounts.

PERSONNEL HAZARDS OF RECYCLING

In 2008, the Waste Management and Remediation Services industry (collections) had a fatality rate of 20.3 fatalities per 100,000 full-time equivalent workers, which is over 5 times the fatality rate for all industries. Nearly 60% of these fatalities were transportation related, while 12% were contacted/struck by, and 8% were exposure to harmful substances or fires/explosions.

Metal scrap recycling, also called secondary metal processing, is a large industry that processes, in the United States alone, 56 million tons of scrap iron and steel (including 10 million tons of scrap automobiles), 1.5 million tons of scrap copper, 2.5 million tons of scrap aluminum, 1.3 million tons of scrap lead, 300,000 tons of scrap zinc, 800,000 tons of scrap stainless steel, and smaller quantities of other metals, on a yearly basis.

Scrap metals, in general, are divided into two basic categories: ferrous and non-ferrous. Ferrous scrap is metal that contains iron, while nonferrous metals are metals that do not contain iron.

Many workers are employed by scrap metal recycling industries. Private, non-ferrous recycling industries in the United States employed approximately 16,000 employees in 2011. (Figures were not available for ferrous recycling industries.) In 2011, those nonferrous recycling industries reported approximately 3,000 injuries and illnesses. The most common causes of illness were poisoning (e.g., lead or cadmium poisoning), disorders associated with repeated trauma, skin diseases for disorders, and respiratory conditions due to inhalation of, or other contact with, toxic agents. Of those injuries and illnesses, 701 cases involved days away from work. The most common types of these injuries were sprains and strains, heat burns, and cuts, lacerations, and punctures (BLS, 2003).

Many companies throughout the country use cardboard compacting machines to reduce the volume of cardboard stored on-site (companies typically store the bundled cardboard until picked up for recycling). The unintended activation of a bailing machine can have catastrophic results. For example, consider the following account:

According to CDC (2011a), on July 31, 1996, a 20-year-old laborer (the victim) at a recycling center died as a result of injuries he received after he was caught between the platen and the top of the bailing chamber door of a vertical-downstroke baling machine. There were no eyewitnesses to the incident. Upon arrival at the work site, a co-worker noticed the victim leaning over the top of the baling chamber. When the victim did not return the co-worker's greeting, the co-worker went to investigate and found that the victim had been caught between the platen and the baling chamber door edge. He notified the local volunteer fire department/emergency medical service which responded in 5 minutes. The victim was removed from the baling machine and transported to a local emergency room where he was pronounced dead. Subsequent examination by investigators revealed that the machine's safety gate interlock had been bypassed, allowing the machine to operate with the gate in the raised position.

(CDC, 2011b)

The hazards (along with controls) that workers in recycling may face are provided below.

HAZARDS AND CONTROLS

- Traffic Safety
- Ergonomics
- Lead
- Machine Guarding

Each of these hazards and controls is briefly discussed below.

TRAFFIC SAFETY IN RECYCLING JOBS[1]

To address one of the safety and health issues in the Waste Collection industry, in December 2003, OSHA published a safety and health information bulletin: *Crushing Hazards Associated with Dumpsters and Rear-Loading Trash Trucks*. The bulletin's purpose was to raise the awareness in the Waste Collection industry of the risks that employees working behind collection trucks faced.

In August 2008, as part of the OSHA and Energy Recovery Council (ERS) Alliance, ERC developed the Hauler Safety Campaign, Safety: Do It for Life. The annual campaign educates public and private waste haulers, municipal and private owners and operators, and facility workers about best practices on tipping floor safety. It also encourages the people who haul and dispose of solid waste to focus on their safety practices and their families' and friends' well-being as they do their jobs.

ERGONOMICS IN RECYCLING JOBS[2]

Workers involved in waste collection may be at risk of developing musculoskeletal disorders (MSDs) from workplace activities that force them to work beyond their physical capacities (i.e., lifting an item that is too heavy, or lifting too often, or working in awkward body postures). MSDs are a serious problem as they can increase the number of employees' lost workdays, increase insurance costs, increase training and staffing costs, and reduce operation efficiency and quality. Improvements in workplace designs, workplace, work postures, weight of materials, and other changes allow workers to work within their physical limits and will likely reduce the number errors, sick days, and injuries and enable workers to be more productive and produce a higher- quality product. Ergonomic improvements are often simple and obvious, such as sorting on elevated tables, the use of simple lifting mechanisms, and rotating workers through different job tasks.

LEAD EXPOSURE IN RECYCLING JOBS[3]

Lead overexposure is one of the most common overexposures found in all industries and is a leading cause of workplace illness. Therefore, OSHA has established the reduction of lead exposure as a high strategic priority.

It is also a major potential public health risk. In general populations, lead may be present in hazardous concentrations in food, water, and air. Sources include paint, urban dust, and folk remedies. Lead poisoning is the leading environmentally induced illness in children. At greatest risk are children under the age of 6 because they are undergoing rapid neurological and physical development.

Workers in the recycling industry are vulnerable to lead exposure as the recycled material, especially electronics or scrap metal, may contain lead. As these recyclables are being crushed, burned, or cut, workers can be exposed to airborne lead.

UNEXPECTED MACHINE START-UP IN RECYCLING JOBS[4]

Lockout/Tagout (LOTO) refers to specific practices and procedures to safeguard employees from the unexpected energization or start-up of machinery and equipment, or the release of hazardous energy during service or maintenance activities.

Again, as mentioned earlier, approximately 3 million workers service equipment and face the greatest risk of injury if lockout/tagout is not properly implemented. Compliance with the lockout/tagout standard prevents an estimated 120 fatalities and 50,000 injuries each year. Workers injured on the job from exposure to hazardous energy lose an average of 24 workdays for recuperation. In a study conducted by the United Auto Workers, 20% of the fatalities (83 of 414) that occurred among their members between 1973 and 1995 were attributed to inadequate hazardous energy control procedures, specifically lockout/tagout procedures.

Scrap metal recycling employers may be required to follow the procedures outlined in OSHA standards at 29 CFR 1910.269(d) or at 29 CFR1910.147.

The following are some of the significant requirements of a Lockout/Tagout procedure required under a Lockout/Tagout program.

- Only authorized employees may lockout or tagout machines or equipment in order to perform servicing or maintenance.
- Lockout devices (locks) and Tagout devices shall not be used for any other purposes and must be used only for controlling energy.
- Lockout and Tagout devices (locks and tags) must identify the name of the worker applying the device.
- All energy sources to equipment must be identified and isolated.

- After the energy is isolated from the machine or equipment, the isolating device(s) must be locked out or tagged out in safe or off position only by the authorized employees.
- Following the application of the lockout or tagout devices to the energy isolating devices, the stored or residual energy must be safely discharged or relieved.
- Prior to starting work on the equipment, the authorized employee shall verify that the equipment is isolated from the energy source, for example, by operating the on/off switch on the machine or equipment.
- Lock and tag must remain on the machine until the work is completed.
- Only the authorized employee who placed the lock and tag must remove his/her lock or tag, unless the employer has a specific procedure as outlined in OSHA's Lockout/Tagout standard.

NOTES

1 Based on information from OSHA's *Green Job Hazards: Waste Management and Recycling: Collection—Traffic Safety.* Accessed 7/11/23 @ www.osha.gov/dep/greenjobs/ recycling_traffic.html.
2 Based on information from OSHA's *Green Job Hazards: Waste Management and recycling: Collection--Ergonomics.* Accessed 7/23/23 @ www.osha.gov/dep/greenjobs/recycling_ eronomics.html.
3 Based on information from OSHA's *Green Job Hazards: Recycling: Scrap Metal Recycling—Lead.* Accessed 7/12/23 @ www.osha.gov/dep/greenjobs/recycling_scrap_ metal_lead.html.
4 Based on information from OSHA's *Green Job Hazards: Recycling: Unexpected Machine Startup.* Accessed 7/13/23 @ www.osha.gov/dep/greenjobs/recycling_scrap_metal_ loto.html.

REFERENCES AND RECOMMENDED READING

Centers for Disease Control (CDC). 2011a. Recycling Center Laborer Crushed in Baling Machine—Tennessee. Accessed 1/11/23 @ www.cdc.gov/niosh/face/IN-house/full9 623.html
Centers for Disease Control (CDC). 2011b. *Organic Waste Recycling Facility.* Accessed 6/11/ 23 @ www.cdc.gov/niosh/face/pdfs
US Bureau of Labor Statistics (BLS). 2003. *Waste Management and Remediation Services.* Accessed 1/11/23 @ www.bls.gov/2003/may/naics3_562000htm
US Environmental Protection Agency (USEPA). (2023). *US EPA Energy Star Program.* Accessed 7/5/23 @ http://energystar.gov/index.cfm?c=thermostats.pr_thermostats

13 Green Energy Waste Streams

INTRODUCTION

A green energy system can only be as green and sustainable as the manner in which wastes are minimized, repurposed, and/or disposed. The United States' increasing interest in green energy systems will create new kinds and new volumes of waste. Green energy systems such as wind turbines and solar photovoltaic panels and batteries are essential to the United States' and the rest of the world's transition from fossil fuel energy to green energy. There are energy demands and by-products associated with production of green technologies, and these systems also produce materials requiring careful end-of-life management to avoid creating new superfund sites and wasting scarce and valuable resources. The following discussion provides a summary of the challenges (as of 2023) the United States and other countries will face in the recycling and proper disposal of these wastes (see Figure 13.1).

Key Term Meaning: Repurposed as used in this book means reusing an existing component or part of the component for a different application.

This discussion will focus on solar panels, lithium-ion-powered electric vehicles, and other appliance battery systems and wind turbines (aka windmills). These systems will be abundant and ever-present in the coming years, and each presents significant challenges as the country considers how to manage its waste stream, or end-of-life cycle. The following discussion also touches on the work that federal and state governments and the sector have done and also on ongoing efforts that look ahead to addressing the challenges of green energy waste and end-of-life issues associated with these sources which has been conducted and sponsored by the USDOE and more specifically by EERE, the office that focuses on research activities. So, this discussion summarizes some of the work done by the USDOE along with published information from other organizations on issues related to green energy waste streams to highlight areas where further research and development are necessary.

SOLAR PANELS WASTE

Solar panels, also known as photovoltaic (PV) modules, present an impending expense to homeowners and businesses responsible for their removal and disposal at the end of their useful lives, as well as a potential environmental hazard. The toxic

DOI: 10.1201/9781003439059-16

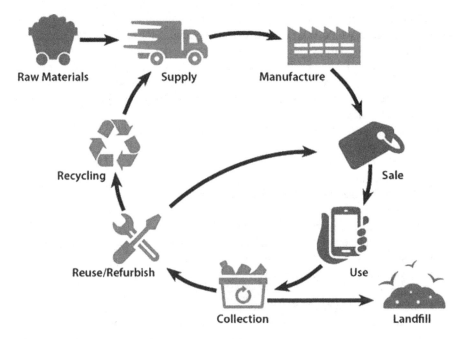

FIGURE 13.1   Life Cycle of Raw Materials.

Source: USGS (2022) Accessed 7/24/23 @ www.usgs.gov/news/national-news-release/US-geological-survey-releases-2022-list-critical-minerals

metals used in PV modules can leach into groundwater if the modules are disposed of improperly. Thus, some PV modules would have the potential to be considered a hazardous waste under the Resource Conservation and Recovery Act (RCRA) because they would fail the test for the toxicity characteristic when the PV modules are no longer in service. At the present time, limited capacity exists to recycle today's solar panels. They are likely to become less valuable to recycle as the manufacture of solar panels becomes cheaper due to a reduction in "critical mineral" usage (see list of critical minerals below). The possible designation of solar panels as RCRA hazardous waste further complicates (can be a real headache-maker) and raises the cost of handling of panels. This increases the prospect of an increasing number of PV modules that are unlikely to be recycled at the end of life. This in turn raises the risk that PV modules will be tossed into municipal landfills and mixed with other household trash, simply discarded by property owners or contractors in places with inadequate controls to restrict open dumping (aka "midnight dumping") (USEPA, 2021).

US GEOLOGICAL CRITICAL MINERALS LIST (2022)

Note: The Energy Act of 2020 defines a "critical mineral" as a non-fuel mineral or mineral material essential to the economy or national security of the United States, and which has a supply chain vulnerable to disruption. Critical minerals are also characterized as serving an essential function in the manufacturing of a product, the absence of which would have significant consequences for the economy or national security. This information provided is essential to understanding mineral dependencies across economic sectors, and to forecasting potential disruptions to mineral commodity supply. As you review this list, notice how many of these critical minerals are related to green energy and green energy practices—batteries, permanent magnets, rare-earth elements, and so forth.

Critical Minerals List

- Aluminum, used in almost all sectors of the economy
- Antimony, used in lead-acid batteries and flame retardants
- Arsenic, used in semiconductors
- Barite, used in hydrocarbon production
- Beryllium, used as an alloying agent in aerospace and defense industries
- Bismuth, used in medical and atomic research
- Cerium, used in catalytic convertors, ceramics, glass, metallurgy, and polishing compounds
- Cesium, used in research and development
- Chromium, used primarily in stainless steel and other alloys
- Cobalt, used in rechargeable batteries and superalloys
- Dysprosium, used in permanent magnets, data storage devices, and lasers
- Erbium, used in fiber optics, optical amplifiers, lasers, and glass colorants
- Europium, used in phosphors and nuclear control rods
- Fluorspar, used in the manufacture of aluminum, cement, steel, gasoline, and fluorine chemicals
- Gadolinium, used in medical imaging, permanent magnets, and steelmaking
- Gallium, used for integrated circuits and optical devices like LEDs
- Germanium, used for fiber optics and night vision applications
- Graphite, used for lubricants, batteries, and fuel cells
- Hafnium, used for nuclear control rods, alloys, and high-temperature ceramics
- Holmium, used in permanent magnets, nuclear control rods, and lasers
- Indium, used in liquid crystal display screens
- Iridium, used as a coating for anodes for electrochemical processes and as a chemical catalyst
- Lanthanum, used to produce catalysts, ceramics, glass, polishing compounds, metallurgy, and batteries
- Lithium, used for rechargeable batteries
- Lutetium, used in scintillators for medical imaging, electronics, and some cancer therapies
- Magnesium, used as an alloy and for reducing metals
- Manganese, used in steelmaking and batteries
- Neodymium, used in permanent magnets, rubber catalysts, and in medical and industrial lasers
- Nickel, used to make stainless steel, superalloys, and rechargeable batteries
- Niobium, used mostly in steel and superalloys

- Palladium, used in catalytic converters and as a catalyst agent
- Platinum, used in catalytic converters
- Praseodymium, used in permanent magnets, batteries, aerospace alloys, ceramics, and colorants
- Rhodium, used in catalytic converters, electrical components, and as a catalyst
- Rubidium, used for research and development in electronics
- Ruthenium, uses as catalysts, as well as electrical contacts and chip resistors in computers
- Samarium, used in permanent magnets, as an absorber in nuclear reactors, and in cancer treatments
- Scandium, used as alloys, ceramics, and fuel cells
- Tantalum, used in electronic components, mostly capacitors and in superalloys
- Tellurium, used in solar cells, thermoelectric devices, and as an alloying additive
- Terbium, used in permanent magnets, fiber optics, lasers, and solid-state devices
- Thulium, used in various metal alloys and in lasers
- Tin, used as protective coatings and alloys for steel
- Titanium, used as a white pigment or metal alloys
- Tungsten, primarily used to make wear-resistant metals
- Vanadium, primarily used as alloying agent for iron and steel
- Ytterbium, used for catalysts, scintillometers, lasers, and metallurgy
- Yttrium, used for ceramic, catalysts, lasers, metallurgy, and phosphors
- Zinc, primarily used in metallurgy to produce galvanized steel
- Zirconium, used in the high-temperature ceramics and corrosion-resistant alloys.

LITHIUM BATTERIES

Question: What is lithium? Lithium, What is it?

Answer: Lithium is the third-place element.

Question: What?

OK, hang on. Let's begin with the basics. In Figure 14.3, the periodic table is shown, and for those of you who are chemists or other scientists, you are very familiar with the table. However, for those who are not that familiar with the table, a brief introduction is called for here and is provided. Note that lithium is listed as 3 out of the 118 elements on the periodic table.

Periodic Table
Lithium (Li) is listed and shown in the periodic table (see Figure 13.2) in the 6 alkali metals—lithium (chemical symbol: Li), sodium (Na), potassium (K), rubidium (Rb), cesium (Cs), and francium (Fr). These are all important chemicals, and without sodium and potassium, we could not live. With the exception of lithium, which is very common, rubidium, cesium, and francium are rare. Lithium, again, is not rare and is used extensively in the batteries that power up small electronic devices including laptop computers, MP3 players, pocket calculators, and several other devices. Lithium is also used as the critical element in EV car batteries.

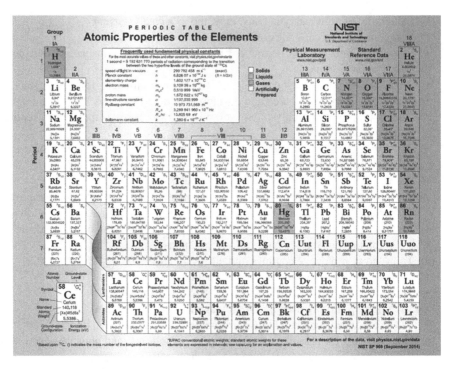

FIGURE 13.2 The periodic table. Lithium (Li) can be found on the far left-hand side of the periodic table in group 1, or 1A, in the alkali metals directly beneath hydrogen (H).

Source: NIST Periodic Table accessed 10/17/2022 @ www.NIST.gov. (US Department of Commerce).

Lithium along with the other alkali metals are found on the far-left side of the periodic table in group 1, or 1A as shown in Figure 13.2. The periodic table is arranged according to increasing atomic numbers. The rows on the periodic table are called periods. Lithium is in period 2. The columns of the periodic table are called groups, or families. Lithium along with the other alkali metals are in group 1.

Note: To gain understanding of the makeup of Li and its electrical properties, it is important to have a basic understanding of matter. The following explanation of matter is given in the sense of electron flow (aka electricity) because it is the electrical aspects of lithium that is the focus of this text. However, the current uses of lithium, including medical uses, are discussed in this text.

ATOMIC STRUCTURE OF LITHIUM

A neutral lithium atom has three protons and three electrons (number 3 out of 118 elements in the periodic table; Figure 13.3 shows a sample of lithium in oil and lithium's chemical symbol). In the periodic table, an element's period, or row,

FIGURE 13.3 Lithium samples stored in oil (for safety) and lithium's chemical symbol indicating atomic number 3 and approximate atomic weight of 7.

Source: F.R. Spellman, 2023. *The Science of Lithium.* **Boca Raton, FL: CRC Press.**

FIGURE 13.3 (Continued)

indicates how many energy levels are occupied by the electrons in its atoms. Lithium is in period 2, indicating the element's three electrons are found on two energy levels. Two electrons are held in the first energy level. The last electron is found on the second, outermost energy level. Note that electrons found on the outermost energy level of an atom are called *valence* electrons. Valence electrons are important because they are the electrons that are involved in chemical bonding with other atoms.

DID YOU KNOW?

Lithium (li) is very reactive, and if unable to sustain normal room temperature, it may catch fire. Kerosene oil is used to ignite fire but is a safe haven for storage of reactive chemicals such as Li.

BASIC PROPERTIES OF LITHIUM

A few important properties of lithium are given in Table 13.1.

LITHIUM COMPOUNDS

Lithium in pure form has limited but important uses, but lithium compounds such as lithium carbonate, lithium stearate, and lithium hydroxide are widely used in everyday life and are briefly described below:

TABLE 13.1
Properties of Lithium

Property	Value
Atomic Mass	6.941 gm
Density	0.535 gm/cm3
Atomic Number	3
Atomic Symbol	Li
Appearance	silvery-white
Atomic Radiance	152 pm
Melting Point	180.5° C (356.9° F)
Boiling Point	1,342° C (2,448° F)
Critical Point	3220K, 67 MPa
Heat of Fusion	3.00 kJ/mol
Heat of Vaporization	136 kJ/mol
Molar Heat Capacity	24.860 j/(mol·K}
Oxidation State	+1
Electronegativity*	0.98 (Pauling Scale)
Electronic Configuration	[He]2s1

*Electronegativity, symbolized Y, is the tendency for an atom of a given chemical element to attract shared electrons while forming a chemical bond. As shown in Table 13.1, lithium's electronegativity is 0.98 on the Pauling Scale. This value, 0.98, is based on calculation because electronegativity can't be measured; it must be calculated.

LITHIUM CARBONATE

Lithium carbonate (Li_2CO_3), only found in the anhydrous state, is the lithium white salt of carbonate and is an inorganic compound widely used in the processing of metal oxides. Lithium carbonate is extracted primarily from spodumene in pegmatite deposits, and lithium salts from underground brine reservoirs. Lithium carbonate is added to glass or ceramic to make the materials stronger. Pyrex™ glass cookware is a good example of where lithium carbonate is used. It is also used in mirrors and lenses for telescopes. Lithium carbonate is also used to extract aluminum metal from aluminum ores. The main use of lithium carbonate is in lithium-ion batteries where its compounds are used as the cathode and the electrolyte. As early as the 1840s, lithium carbonate was used to treat stones in the bladder, and by the 1850s, lithium carbonate salts were used to treat gout, urinary calculi, rheumatism, mania, depression, and headache. By the 1940s, lithium carbonate became a common treatment for bipolar disorder.

LITHIUM STEARATE

Lithium stearate ($LiC_{18}H_{35}O_3$) is a white solid powder that is mixed with petroleum to make a heavy, lubricated grease used in many industrial machines; it is also used in making soaps. The grease does not break down at high temperatures and does not get hard at low temperatures. It also does not react with oxygen in air or in water, so the chemical composition has a long use life.

LITHIUM HYDROXIDE

Lithium hydroxide (LiOH) is an organic compound that can exist as anhydrous or hydrated and that is useful in absorbing carbon dioxide.

DID YOU KNOW?

Because demand is soaring, lithium is often referred to as "the new oil" and also as "the new white gold" because it is seen as a valuable alkali metal that is able to store huge amounts of energy, all of it squeezed into a very small area, space, and/or container—such as a storage battery. The demand for lithium is soaring and competition to find and produce lithium is significant and in some cases is filled with drama.

LITHIUM-ION BATTERY WASTE

Lithium is one of many elements and compounds used in several electronic devices. Figure 13.4 shows several elements, minerals, or compounds that are used to manufacture an iPhone. Lithium is necessary for a number of things including strategic, consumer, and commercial applications. The primary uses for lithium are in batteries (*Note:* an extensive discussion of lithium batteries is presented later in this text), ceramics glass, metallurgy, pharmaceuticals, and polymers.

The huge plus, the advantage, and the benefit of lithium use in electronic devices is its excellent electrical conductivity and very low density that allows it to float on water. These attributes make lithium an ideal component, for example, for battery manufacturing. Note that lithium is traded in three different forms: mineral compounds from brines, mineral concentrates, and refined metal via electrolysis from lithium chloride. It is important to point out that lithium minerology is diverse; it is found in a variety of pegmatite minerals such as spodumene, lepidolite, amblygonite, and in the clay mineral hectorite. At the present time (2023), global production of lithium is dominated by pegmatite and closed-base brine deposits, but there are significant resources in lithium-bearing clay minerals, oilfield brines, and geotherm brines (Bradley et al., 2017).

Here's the 411 on lithium-ion battery waste: those lithium-ion batteries (LIBs) that no longer hold enough power for automotive use will pose problems even before we have a wave of problems with solar waste. Lithium-based electric car batteries have caused fires and typically contain toxic metals—both problems make recycling and/or disposal more complicated.

The problem is that there is more than just fires in automobiles—a real head-scratcher or worse—LIBs have caused a growing number of fires in the waste

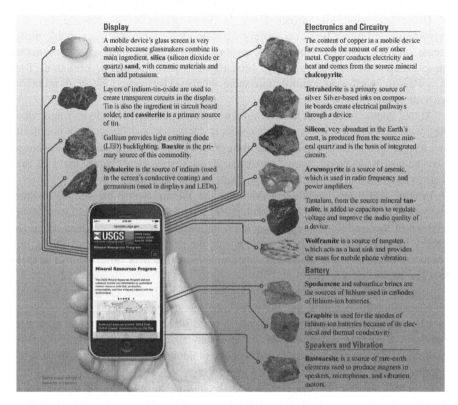

FIGURE 13.4 Shows the various minerals/chemicals used for an iPhone. Public Domain photo from USGS (2018). Accessed 1/28/23 @ www/usgs.gov/data/lithium-deposits-united-states/

management process. The USEPA (2021) points out that anecdotal information has shown that materials recovery facilities (i.e., recycling centers, or "MRFs") and other waste facilities have seen an increased number of fires due to LIBs, but, at the present time, there has been limited data on fires incidents at a national level. One of the problems is that not all fires caused by improperly discarded LIBs are captured and not all incidents are covered by the news. Due to these limitations, the relatively easy way to collect data from environmental organizations was employed herein to gather representative data. Keep in mind that it is likely that a significant number of additional relevant cases have not been identified.

The USEPA (2021) found that 64 waste facilities experienced fires 245 fires that were caused by, or likely caused by, lithium metal or lithium-ion batteries. Note that these facilities included MRFs, transportation vehicles (garbage truck, etc.), landfills, and other waste management industry locations (e.g., electronics recycles, transfer stations, etc.). These fires occurred between 2013 and 2020 in 28 states and in all 10 EPA Regions.

These fires not only affected the facilities where they occurred but also surrounding communities in a variety of ways, which include four main impact areas: injuries, external emergency response, service disruptions, and monetary losses. Note that some fires caused little to no impact, such as several small fires at a Pacific Northwest landfill that were easily extinguished by staff members without issue.

At this Pacific Northwest landfill, fires were not a problem 10–15 years ago, when they were less common in consumer products. However, as LIBs grew in popularity and usage, the landfill started seeing more fires caused by these batteries. Although LIBs ae now widespread, many consumers do not realize the difficulty that LIB fires pose to landfills and continue to discard batteries in the trash.

In June 2017, the landfill supervisor in the Pacific Northwest started tracking every fire caused by LIBs in the landfill. Every time an employee pulled a smoldering LIB out of the landfill, the supervisor kept meticulous records of the date, and if the battery had not been burned past recognition, the source of the battery (tablets, cell phones, computers, etc.). These LIB fires have increased over time, from 21 in 2018 to 47 in 2020 (USEPA, 2021).

It turned out that the most common sources of LIBs that caught fire were cell phones, followed by tablets and laptops. Although all the recorded fires were caused by LIBs, the specific type of battery was often difficult to ascertain, with most fires caused by LIBs from unknown sources. The truth be known, when a battery burns for too long, it can become charred (well cooked) enough that it is impossible to determine what type of device it originated from.

Note that most of the fires caused by LIBs at this facility occur on the surface of the landfill. What this means is that the fires are often easy to spot, and employees usually tackle them right away. What site employees have been trained to do is to make certain that every vehicle that goes out onto the landfill has a fire extinguisher. When a fire starts, a worker douses the area, recovers the device, puts it in a steel bucket, and sends the LIB to a hazardous waste facility.

The fly in the ointment in this procedure is that the site is not manned 24/7, but when manned within business hours, employees are present to extinguish and properly dispose of the LIBs. However, at night, fires have the chance to grow large, beyond the surveillance of landfill employees. This facility had two such larger fires in May 2019 and May 2020 during off-hours and required the help of the local fire department.

The landfill operator has gone beyond only tracking incidents and now also educates consumers about the dangers of LIBs. Sometimes, large quantities of returned electronics from retail stores end up at the landfill. These goods can't be sold, and store owners incorrectly believe that electronics with batteries can be thrown away in the trash. When this occurs, the landfill manager reminds stores that these items should not be discarded with their batteries inside. The landfill manager has also educated the public by participating in local radio and television news segments, where he explains the different types of batteries and the damage they can cause to a landfill. The manager hopes that getting the word out on LIBs will reduce the number of fires at landfills and other waste management facilities.

The manager assumes other landfills are most likely seeing the same frequency of fires as his facility sees. Because LIB fires are often small and unreported at this landfill, this is probably like other small fires at other landfills. Usually these landfill LIB fires are routine and not newsworthy, and they are easy to put out if noticed immediately.

Let's look at another industry experience with LIB fires.

At Larimer County landfill's recycling center in Fort Collins, Colorado, a fire broke out in August 2018 that fire inspectors believe was ignited by an LIB. The good news is that an experienced operator noticed the fire and used processing equipment to move the burning pile of recyclables out of the facility. This quick thinking likely saved Larimer County from experiencing many of the detrimental effects often caused by LIB-induced fires. The fire department arrived and was able to subdue the blaze outside the building, saving the valuable equipment inside and sparing the structure from damage. The truth be told, the fire department indicated that the entire facility likely would have been lost if not for the quick response by the operator. Note that the facility also didn't see any increased insurance rates from this fire; it is self-insured by the county.

This does not mean that the facility did not experience some undesirable outcomes, though they were mitigated. Multiple employees were treated at the scene for smoke inhalation, but fortunately none had to be hospitalized. Additionally, the facility was forced to shut down for the rest of the day, temporarily disrupting the county's recycling collection service.

The bottom line: improper disposal of LIBs can lead to impacts on workers and the local community.

OK, flash back to 2018 again where the 2018 blaze was just the beginning of LIB fires experienced at the Larimer County site, often due to LIBs. A representative from the county indicated that they feel the frequency of such fires has been increasing to the point that their landfill sees multiple fires per week and their recycling facility experiences a few per month. Moreover, the county's solid waste department staff fear that this issue, and the threats, dangers, risks, and menaces associated with it, will intensify as the county plans to close their landfill and transition to using a transfer station. There process conditions may increase the frequency of fires, while the facility's characteristics (i.e., being inside one of these structures) may exacerbate the negative impacts if one of these fires gets out of control.

Because of the wake-up call of the near miss of the 2018 fire and the general trend of increasing risk, Larimer Country decided to adopt a multifaceted consumer education campaign to respond to the threat of LIB-caused fires in their facilities. They started a battery safety campaign focusing on raising the community's awareness about the importance of proper disposal of batteries of all types. So, in late 2020, the county solid waste department invited high school- and college-age graphic design students to submit potential designs for the graphic components of the campaign's promotional and information materials. The intent of this contest was to get younger generations involved in the issue and create a vested interest in addressing it. Moreover, Larimer County is working with a student intern to develop an educational presentation and video for the county's website.

The ultimate success of Larimer County's efforts is as of yet unknown, but their experiences overall illustrate the challenges that local waste management authorities are facing and the innovative strategies some of them are adopting to attempt to combat this nationwide issue (USEPA, 2021).

The takeaway from both examples, like Larimer County's experiences with LIB-caused fires, boils down to the difficulty of separating the component materials from one another in an economical manner—a key obstacle to recycling these batteries. At present, manufacturers are taking a few actions so that the burden does not fall directly on EV owners. Forms of adaptive reuse are underway, sponsored by certain automakers and battery producers, but they only slightly delay the inevitable: a time when the batteries no longer have value as a battery and require recycling or disposal. It is important that these batteries are properly collected and managed to eliminate the rise of fires or the leaching of metals into the environment, creating future Superfund sites (USEPA, 2021).

DID YOU KNOW?

Billowing smoke quickly killed four people asleep in apartments in New York City in June 2023. The apartments were located above a bike shop. Early on that June morning, an explosion occurred in the bike shop and ignited a blaze that engulfed the building. The shop fire was caused by exploding lithium batteries (or a single battery) for motorized bicycles. The truth be known, with the popularity of e-bikes growing, so has the frequency of fires and deaths blamed on the batteries that power the bicycles. These fire events have prompted a campaign to establish regulations on how the batteries are manufactured, sold, reconditioned, charged, and stored (Calvan, 2023).

SIDEBAR 13.1 LITHIUM-ION VERSUS SODIUM ION BATTERIES

Few people are likely to dispute that batteries are becoming crucial to everyday life. If you want to win a lottery, produce a better battery that will collect for you an untold fortune, worldwide fame spiced with some headaches (legal implications like being sued for some assumed or alleged safety and health problem with your product—remember, in the present climate the mantra seems to be: Let me sue you before you sue me). Note that sodium-ion batteries are a top candidate for the crown held by lithium-ion batteries—the new White Knight.

The new White Knight? How so?

Well, it is so in several ways. First, the sodium-ion (Na-ion) battery is the new White Knight (potentially—a savior in emerging battery technology) in battery technological advancement since the 1990s and might be a possible alternative to lithium-ion batteries, the type commonly found in EVs and phones.

So, how do sodium-ion (Na-ion) batteries work?

In the simplest terms possible, it can be said that Na-ion batteries use a chemical reaction to store and release electrical. As is the case with all batteries, they have a positive and a negative electrode. The electrodes are separated by an electrolyte—a special substance that allows tiny particles (ions) with a positive or negative charge to move between the electrodes. Na-ion batteries work similarly to lithium-ion batteries (currently this battery type is the rooster in the hen house, so to speak), but instead sodium-ion batteries use sodium ions and not lithium ions. The negative electrode (the cathode) of the Na-ion and the electrolyte contain sodium. Plainly, the performance and life span of a battery is important. Therefore, the hunt is on for the best electrolyte to use because of its effect on the performance and life span of the battery. We continue to search for combinations (continuously, we hope) that give the best performance, safety, and, mainly, availability and cost (although the goal should be safety and health first and always).

Okay, what makes a Na-ion so great?

Another good question. While it is true that lithium-ion batteries are the rooster in the hen house (they rule the roost, so to speak), it is important to note that the Na-ion battery has a few distinct advantages over the so-called rooster ruling the roost. Let's look at the advantages of Na-ion batteries. Note that like about anything else, there are advantages and the disadvantages with Na-ion— and both are listed below.

Advantages:

- Sodium, chemical symbol Na, and atomic number 11 is a soft, silvery-white, highly reactive metal that is a much more abundant element than lithium. Sodium is easy and cheaper to obtain. This makes Na-ion batteries less expensive to manufacture than lithium-ion batteries.
- Na-ion batteries are more environmentally friendly than non-Na-ion batteries.
- Na-ion batteries can be tooled to provide similar energy density as LIB batteries. This is important because it makes the Na-ion battery suitable for a wide range of applications.
- Na-ion batteries are safer than LIB batteries. There have been incidents where LIB batteries have caught on fire. EV owners must worry about the potential for battery fires when involved in crashes. Moreover, EV automobiles have generated a few fires while inside residents' garages. The point is that Na-ion batteries are safer than LIB batteries.

Drawbacks:

- Na-ion batteries have a 2.5 voltage rating, while the LIBs deliver 3.7 volts. What this means is that Na-ion batteries may be unsuitable for systems or components that need high power and require a lot of energy to be delivered quickly.
- Na-ion batteries have a slower charge/discharge rate than LIBs and may be unsuitable for providing quick power to EVs.
- Presently, Na-ion batteries have limited charge cycles before battery degradation; some LIB batteries can reach about 10,000 cycles before degradation.

Apart from the technical pros and cons, Na-ion batteries are a work in progress. Researchers, engineers, and others are working on solutions to its remaining weak points—this includes improving their performance and commercial viability. At the present time, the focus is on improving performance by improving the energy density and voltage of Na-ion batteries and their life span charge/discharge rate.

The bottom line, at least at the present time, is that LIBs provide superior performance compared to sodium-ion batteries. But this may change with advancements in Na-ion batteries.

WIND TURBINE WASTE

Depending upon where you live, wind turbines and the energy they deliver is one of the fastest growing sources of electricity generation growth and is vital to the goal of green energy production, that is, in turn, vital to reducing greenhouse gas emissions. One of the problems with green wind energy production is that the manufacture of wind turbines requires large quantities of materials and components. Another problem with wind turbines is that these huge power generators are not only laborious and expensive to transport and install, but they are just as burdensome to remove and recycle. The point is, waste management problems and challenges associated with wind turbines are due to their size. The biggest difficulty in their storage and disposal is that they are large structures composed of large components and due to their mostly fiberglass composite construction, are of limited material reuse value. A temporary solution to this problem is the shipping of outdated wind turbine units to less developed countries. However, this solution does not eliminate the need to appropriately manage wind turbines at the end of their useful lives. What is really happening with this practice is the passing off and dumping of end-of-useful-life turbines and components to countries ill-equipped to properly dispose of them. While it is true that the burden of management falls not on the landowner but on the owner of the wind turbine, they may also pose a threat to groundwater if transformer fluids and lubricants used by wind turbines leak or are dumped and contaminate land.

THE BOTTOM LINE

Using, reusing, recycling, and remanufacturing solar, EV battery, and other battery applications, and wind turbine materials will reduce waste and create a "circular economy."

A circular economy for energy materials is the subject of the final chapter.

REFERENCES

Bradley, D.C., Stillings, L.L., Jaskula, B.W., Munk, L.A., and McCauley, A.D. 2017. Lithium, chapter. In Schulz, K., DeYoung, J.H., Jr., Seal, R.R., II, and Bradley, D.C., eds., Critical mineral resources of the United States—Economic and environmental geology and prospects for future supply. US *Geological Survey Professional Paper* 1802, p. K1–21. https://dol.or10.3133/pp1802K

Calvan, B.C. 2023. E-bike battery fires on rise. The Virginian-Pilot, July 29.

US Environmental Protection Agency (USEPA). 2021. *Renewable Energy Waste Streams.* Accessed 7/25/23 @ www.epa.gov/newsreleases/epa-releases-briefing-paper-renewable-energy-waste-management

14 Green Energy in a Circular Economy

THE 411 ON CIRCULAR ECONOMY

OK, to begin with a circular economy for energy materials means that technology should be engineered from the state to require few materials, resources, and energy while lasing longer and having components that can easily be broken down for use in subsequent applications. For the purposes of this book "Circular Economy" is defined as an economy that keeps materials, products, and services in circulation for as long as possible (USAID, 2023). The Save Our Seas 2.0 Act, for example (addresses the increasing quantities of plastic waste in the seas), refers to an economy that uses a systems-focused approach and involves industrial processes and economic actions that are restorative or regenerative by design, enables resources used in such processes and activities to maintain their highest value for as long as possible, and points toward the elimination of waste through the superior design of materials, products, and systems.

DID YOU KNOW?

In the circular economy, an arbitrage opportunity entails the benefits in terms of material costs, labor, and energy that circular setups provide over linear models (WEF, 2014).

Simply, a circular economy shrinks material use; redesigns materials, products, and services to be less resource intensive; and recaptures "waste" as a resource to manufacture new materials and products.

What we are talking about here is a paradigm shift whereby there is a change to the model in which resources are mined, made into products, and then become waste. Let's state this one again, differently, to enhance understanding of a circular economy—the goal of the circular economy is to transition from today's take-make-waste linear pattern of production (see Figure 14.1) and consumption to a circular system in which the societal value of products, materials, and resources is maximized over time. However, make a note of the fact that circularity in and of itself does

DOI: 10.1201/9781003439059-17

not ensure social, economic, and environmental performance (i.e., sustainability). Sustainability of circular economy strategies needs to be evaluated against their linear (direct) counterparts to identify and avoid strategies that increase circularity yet lead to unintended outwardness. While it is true that the proliferation of circularity metrics has received considerable attention, at the present time, there is no critical review of the methods and combinations of methods that underlie those metrics and that specifically quantify sustainability impacts of circular strategies (EERE, 2023).

DID YOU KNOW?

End-of-use refers to materials/products at the end of their primary use, that are collected and returned to the same usage, or cascaded to a new one (WEF, 2014).

DID YOU KNOW?

It is important to note that the circular economy is not the linear economy that most of us are familiar with, or have been living with and become used to many of us without even knowing or understanding the approach that has been practiced since the Industrial Revolution in the eighteenth century. The linear economy involves taking raw materials from the environment and turning them into finished product. Then the finished products are used and then discarded into the environment. And in many cases the old "I do not want this product anymore so just throw it into the river, lake, ocean, or landfill—out of sight, out of mind syndrome." The problem with this linear economy should be rather obvious because this system has a beginning and an end where limited raw material, critical minerals, run out in the long run. So, what we end up with in this linear economy is a pile of waste, growing by the minute, bringing about more waste-disposal issues and expenses.

What the shift from a linear to a circular economy requires is for us to make products differently. Considering this shift, the life of materials must be extended beyond their original intended use and not subject to being landfilled when damaged, worn out, and no longer viable for any further use. Note that this is more than redoubling efforts on recycling. Circular economy strategies that reduce the use of raw materials can significantly cut global greenhouse gas emissions and relieve pressure on raw materials such as metals (Circularity Gap, 2020). The idea is to strive to reach a pathway to zero economy and to sustain economic growth.

Shifting to green energy complemented by energy efficiency cuts 55% of global greenhouse gas emissions (Ellen MacArthur Foundation, 2019). Keep in mind that for green energy to be a truly clean power source, solar, wind, and batteries and accessories must be fashioned, utilized, retired in a net-zero, safe, and sustainable

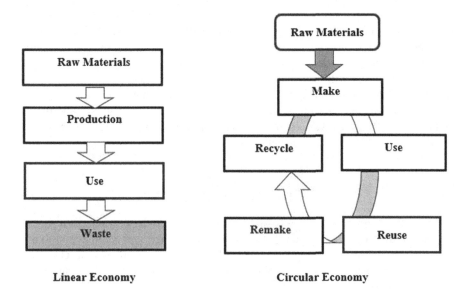

FIGURE 14.1   Linear and circular economy.

Source: Adaptation from C. Westman (2016). Accessed 7/27/23 @ https://commons.
wikimedia.org/wiki/file:linear.com/

way. A point of concern is that at the current rates of production, the annual need for critical materials, especially critical metals, could endanger the transition to a clean energy economy. Regarding the reference to critical metals, two such examples include the metals indium and silver that are needed for solar panels, and neodymium that is used in magnets in both wind turbines and EVs. When mined, these and other resources produce contaminated waste, contribute to biodiversity loss and ecological degradation, and involve frequent human rights abuses. Regarding human rights abuses and their connection to mining activities, they include increased road traffic, access to water and impacts of poor-quality water, increases in noise and air pollution, and child labor involved in the global mining for lithium, cobalt, manganese, copper, nickel, and zinc—all used in green energy technologies like wind turbines, EVs, and solar panels.

The other fly in the ointment is that a significant amount of green energy equipment is approaching the end of its useful life; the quantity of obsolete green energy equipment is expected to grow exponentially over the next three decades. For example, consider that in 2019, the world generated 18,000 tons of solar photovoltaic (PV) panel waste. It is estimated that by 2050, PV panel waste could increase to 10 million tons annually (BNEF, 2020). Wind Europe (2019) estimates that around 14,000 wind turbine blades could be mothballed by 2023, equivalent to between 40,000 and 60,000 tons. In addition, there is a concern with and over what we call and classify as e-waste (electronic waste), or separately, concerns about unsafe handling of waste that results in harm to human health and the environment. By returning this equipment to be reused or recycled, circular economy strategies both keep waste out of landfills and moderate demand for the materials that make energy equipment to begin with.

According to Accenture (2015) research, the circular economy could generate trillions of dollars in economic output by 2030; moreover, circular business models were identified that will help uncouple economic growth and natural resource consumption while driving competitiveness. The circular economy focuses on reuse, repair, remanufacture, and the sharing of other significant innovation opportunities and jobs. By 2050, more sustainable use of materials and energy could add more than $1 trillion annually to the global economy.

DID YOU KNOW?

Reshaping the Green Energy Industry into a more circular economy is a paradigm shift that has significant potential to reduce waste and carbon emissions while extending the supply of valuable resources. Moreover, a circular economy also cultivates new business models, innovations, initiatives, sustainable practices, policies, and markets.

THE TRANSITION: LINEAR TO CIRCULAR ECONOMY

Take + Make + Dispose = Linear Economy
Returning By-products + Circular Supplies + Waste as a Resource = Circular Economy
Repairing, refurbishing, and reusing = Circular Economy

In a linear economy where tons of materials are extracted and processed, contributing to at least half of global carbon dioxide emissions, the resulting waste—including plastics, textiles, food, electronics, and more—is taking its toll on the environment and human health. On the other hand, and in contrast to a linear economy, a circular economy basically circulates materials and products instead of producing new ones (see Figure 14.1). Moreover, whereas the linear economy disposes of waste in landfills at the end of its use (or when we do not want it anymore), the circular economy creates several different opportunities for return cycles, or what are known as loops that avoid disposal or landfilling. As mentioned, this closed-loop system minimizes the use of resource inputs and the creation of waste, pollution, and carbon emissions by keeping materials, products, equipment, and infrastructure in use for longer periods, thus improving the productivity of these resources.

DID YOU KNOW?

According to USAID (2023), in a circular economy, recycling becomes the last resort, not the first and only option.

CHINA'S LINEAR ECONOMY

The linear economy's use of critical metals depends primarily on China, a country that has threatened to disrupt the green energy supply chain and that has a record of human rights abuses and very poor environmental practices (Bemreuter, n.d. and Sheffield Hallam University, n.d.). Polycrystalline silicon (aka multicrystalline silicon, polysilicon, poly-Si or mc-Si) is a high purity form of silicon, used as raw material by the solar photovoltaic and electronics industry made in China. China is also the main source of tellurium (a chemical element that is brittle, mildly toxic, rare, and silver-white; it is an allotrope meaning it exists in two forms: crystalline and amorphous like diamond and graphite, which are two allotropes of carbon). China is also the main source of indium (a chemical element with the symbol In and the atomic number 49), another critical metal like tellurium, both of which are used create some thin-film technologies for solar PV modules. China and Russia are the source for the neodymium, dysprosium, praseodymium (see Figures 14.2 and 14.3), and terbium components of permanent magnets used in wind turbines and EVs. These four components are so important to and in this discussion of green energy that they are described in some detail in the following sections.

FIGURE 14.2 Chemical element designation for praseodymium with chemical symbol, element name, atomic number, and atomic weight.

$$\text{Xe } 4f^3 \, 6s^2$$

FIGURE 14.3 Illustrates praseodymium's electron configuration.

PRASEODYMIUM

Details:

Melting point—1208 K (3130° C, 5666° F)
Phase at standard temperature and pressure (STP)—solid
Appearance—grayish-white
Boiling point—3130° C
Density—6.77 g/cm^3
Solubility in water—decomposes
Heat of fusion—6.89 kJ/mol
Vickers hardness—210-470 MPa
Heat of vaporization—331 kJ/mol
Electronegativity—1.13
Natural occurrence—primordial
Magnetic ordering—paramagnetic
Naming—from the Greek meaning "green twin"
Curie point—NA

Praseodymium, Pr, is a silvery, soft, malleable, and ductile metal (with hardness comparable to silver), prized for magnetic, electrical, chemical, and optical properties, and many of its industrial uses involve its ability to filter yellow light from light sources. Praseodymium is too reactive to be found in natural form; when in pure form and exposed in air, it develops a green oxide coating. Praseodymium applications include use as magnets, lasers, pigments, and cryogenic refrigerants. More specifically, praseodymium in magnet use finds applications in small equipment, such as printers, watches, motors, headphones, loudspeakers, and magnetic storage (McGill, 2021). Praseodymium compounds give enamels, glasses, and ceramics a yellow color (Lide, 2007; McGill, 2021).

DYSPROSIUM

Details:

Figure 14.4 illustrates dysprosium's chemical symbol, element name, atomic number, and atomic weight; Figure 14.5 illustrates dysprosium's electron configuration.

Melting point—1680 K (1407° C, 2565° F)
Phase at standard temperature and pressure (STP)—solid
Appearance—silvery-white
Boiling point—2562° C
Density—8.540 g/cm^3
Solubility in water—insoluble
Heat of fusion—11.06 kJ/mol
Vickers hardness—410-550 MPa
Heat of vaporization—280 kJ/mol

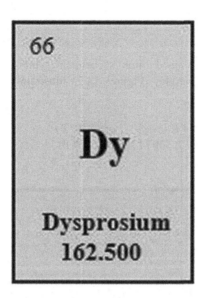

FIGURE 14.4   Chemical element designation for dysprosium with chemical symbol, element name, atomic number and atomic weight.

$$Xe\ 4f^{10}\ 6s^2$$

FIGURE 14.5   Dysprosium electron shell configuration.

Electronegativity—1.22
Natural occurrence—primordial
Magnetic ordering—paramagnetic at 300 K
Naming—after the Greek, meaning "hard to get"
Curie point—87 K

Dysprosium, Dy, with a metallic silver luster was first isolated in the 1950s after development of an ion-exchange technique; it is never found in nature as a free element but instead in the mineral xenotime (a rare-earth phosphate mineral), and in other minerals. Naturally occurring dysprosium is composed of isotopes; the most abundant is [164]Dy. In usage, dysprosium has very few uses or applications because it can be replaced by other chemical elements. When used, it functions as a high thermal neutron absorber in control rods for nuclear reactors or in data-storage applications because it has high magnetic susceptibility, it is also is used as a component of Terfenol-D (it expands and contracts in a magnetic field). With regard to toxicity, Dy is mildly toxic when in soluble form but not toxic in insoluble salt form. Caution is advised whenever Dy is in powder form because it may present an explosion hazard. A Dy fire cannot be extinguished with water because it reacts with the water to produce flammable gas (Dierkes, 2003).

NEODYMIUM

Details:

Figure 14.6 illustrates neodymium's chemical symbol, element name, atomic number, and atomic weight; Figure 14.7 illustrates neodymium's electron configuration.

- melting point—1294 K (1021° C or 1870° F)
- boiling point—3347 K (3074° C or 5565° F)

60
Nd
Neodymium
144.242

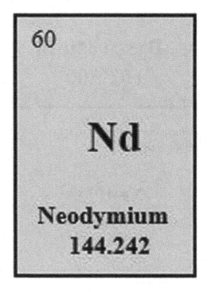

FIGURE 14.6 Chemical element designation for Neodymium with chemical symbol, element name, atomic number, and atomic weight.

$1s^2$
$2s^2\ 2p^6$
$3s^2\ 3p^6\ 3d^{10}$
$4s^2\ 4p^6\ 4d^{10}\ 4f^4$
$5s^2\ 5p^6$
$6s^2$

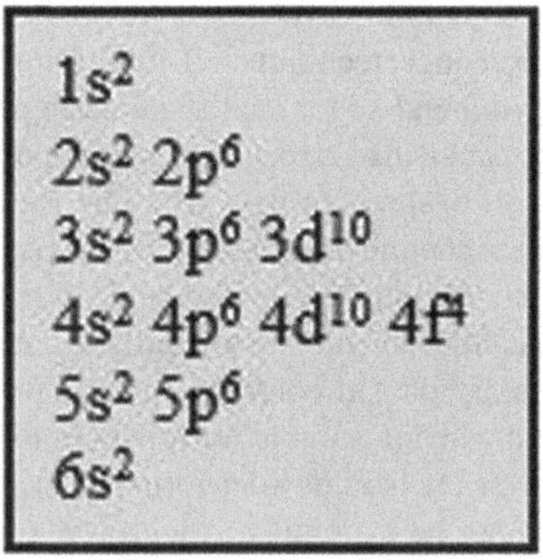

FIGURE 14.7 Neodymium electron shell configuration.

- density—7.01 grams per cubic centimeter
- solid at room temperature
- element classification—metal
- periodic element—6
- group name—lanthanide
- electronegativity—Pauling scale: 1.14

Note: In the simplest terms I can render, *electronegativity* is a measure of the ability of an atom to attract the electrons when the atom is part of a compound. When going from left to right across the periodic table, electronegativity values generally increase. However, when going from top to bottom of a group, electronegativity values decrease. The highest electronegativity value is for fluorine, at 4.0, and the lowest is for cesium, at 0.79. Moreover, as stated above, the electronegativity value for neodymium is 1.14 on the Pauling scale.

- natural occurrence—primordial (i.e., from the beginning of time)
- crystal structure—double hexagonal close-packed
- name—from the Greek *neos* and *didymos*, meaning "new twin"
- pronounced—nee-eh-DIM-ee-em

DID YOU KNOW?

Nd glass is produced by the inclusion of neodymium oxide (Nd_2O_3) in the glass melt. When this is accomplished and the glass is exposed to daylight or incandescent light, the glass appears lavender in color, but under fluorescent light neodymium glass appears pale blue. The color scheme can be changed or altered to appear ranging from pure violet through wine-red or warm gray.

The point Cascio (2023) makes is correct about neodymium being one of 17 rare-earth metals that is critical to the green technology industries. Neodymium is a powerful magnetic material. An Austrian chemist, Carl E. Auer von Welsbach, discovered neodymium in 1885. From a material known as didymium he separated both neodymium and the element praseodymium. At the present time, neodymium, primarily a reddish-brown phosphate mineral occurring usually in small, isolated crystals, is primarily obtained via an ion exchange process monazite sand ((Ce, La, Th, Nd, Y)PO_4), a material rich in rare-earth elements; lanthanum (La) is the most common rare-earth element in monazite. Monazite is radioactive due to the presence of thorium and in lesser amounts, on occasion, uranium.

Neodymium has a variety of present uses. For example, approximately 18% of Misch metal (aka Mischmetal from German: *Mischmetall*—"mixed metal") is neodymium that when mixed and hardened with iron oxide and magnesium oxide, makes ignition devices in pyrophoric ferrocerium devices—misnamed "flint," but is used to produce hot sparks that can reach 5,430° F (3,000° C) which are used as strikers for gas welding, cutting torches, and so forth.

Another important use of neodymium, beginning in 1927 and still popular today is its usage as glass dyes. Neodymium compounds often are of a reddish to purple color; however, it changes with the type of lighting exposure because it contains sharp light absorption bands intermixed with emission bands of mercury and other elements. When glasses are mixed or doped with neodymium, they can be used, and are used, in infrared lasers with wavelengths ranging from 1,047 to 1,062 nanometers. These are used in experiments in inertial confinement fusion (ICF)—which initiates nuclear fusion by compressing and heating targets filled with thermonuclear fuel. Neodymium has an unusually large specific heat capacity (i.e., the heat required to raise the temperature of a unit mass of substance by a given amount) at liquid-helium temperatures, so it is useful in cryocoolers (a refrigerator designed to operate at 123K, which equals −150° C or −238° F).

Neodymium is also used as a component in alloys employed to make high-strength neodymium magnets. At the present time, neodymium permanent magnet (PM) generators are the trend. The widespread practice today is to use rare-earth magnets in wind turbines, especially in offshore wind turbines, as they allow for high-power density and diminished size (low mass), and relatively low weight with peak efficiency at all speeds, offering a high annual production of energy with low lifetime expenditures. Most direct-drive turbines are equipped with permanent magnet generators that typically contain neodymium, and smaller quantities of dysprosium. Although on a different extent or scale, the same is true for numerous gearbox designs. Using a straightforward structure, the PMs are installed on the rotor to generate a constant magnetic field. The produced electricity is collected from the stator by using the commutator, slip rings, or brushes. To lower cost, the PMs are integrated into a cylindrical cast aluminum rotor. Note that for onshore wind turbine installations, it is not necessary to utilize permanent magnet generators because reduced size and weight are not a concern (as they are with offshore installations). PM generators are similar in operation to synchronous generators except that PM generators can be operated asynchronously (i.e., induction generators that require the stator to be magnetized from the electric grid before it works). Note that the high-power-versus-weight off-shore wind turbine generators are also used in electric motors for hybrid cars and generators for aircraft.[1] The strong, versatile permanents are widely used in high-performance hobby items, professional loudspeakers, in-ear headphones, and computer hard disks—all these items require high-power, strong magnetic fields where low mass is desired.

Figure 14.8 shows a larger neodymium magnet attracting smaller neodymium-iron-born refrigerator magnets and a few other friends. You really can't appreciate how strong these permanent magnets are until you attempt to pull them apart—no easy task.

Neodymium permanent magnets used as drive motors in some vehicles require about 2.2 pounds (one kilogram) of neodymium per vehicle (Gorman, 2009). It is interesting to note that researchers using high-precision imaging at Radboud and Uppsala Universities observed self-induced spin glass—a behavior in the crystal structure of neodymium due to incredibly small changes in the magnetic structure (Kamber et al., 2020; Radboud University, 2020).

FIGURE 14.8   Neodymium permanent magnet in the process of attracting friends. Photo by F. Spellman.

TERBIUM

Details:

Figure 14.9 illustrates gadolinium's chemical symbol, element name, atomic number, and atomic weight; Figure 14.10 illustrates gadolinium's electron configuration.

Melting point—1629 K (1356° C, 2473° F)
Phase at standard temperature and pressure (STP)—solid
Appearance—silvery white
Boiling point—3396° C
Density—8.23 g/cm³
Solubility in water—insoluble
Heat of fusion—10.15 kJ/mol
Vickers hardness—510-950 MPa
Heat of vaporization—391 kJ/mol
Electronegativity—1.2
Natural occurrence—primordial
Magnetic ordering—paramagnetic at 300 K
Naming—after Ytterby (Sweden), where it was mined
Curie point—222 K

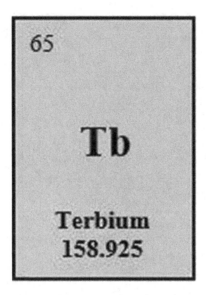

FIGURE 14.9 Chemical element designation for terbium with chemical symbol, element name, atomic number, and atomic weight.

$$Xe\ 4f^9\ 6s^2$$

FIGURE 14.10 Terbium electron shell configuration.

Terbium, Tb, first isolated in 1843, is a silvery-white rare-earth metal and is the ninth member of the lanthanide series. It is malleable, ductile, and soft enough to be easily cut with a knife. It reacts with water and is electropositive. Terbium is not a free element found in nature but is found in many minerals, including cerite, gadolinite, euxenite, xenotime, and monazite. Terbium is used as a crystal stabilizer of fuel cells, which operate at elevated temperatures, and is also used to dope calcium tungstate, calcium fluoride, and strontium molybdate. As a key component of Terfenol-D that is named after terbium, iron (Fe), and dysprosium ($Tb_xDy_{1-x}Fe_2$—x = 0.3), the alloy has the highest magnetostriction of any alloy, up to 0.002 m/m at saturation—it expands and contracts in a magnetic field, making it valuable and essential in the use of actuators, in naval sonar systems, and in sensors. Terbium oxide is used in green phosphors and fluorescent lamps, cathode ray tubes, optical computer memories, and television tubes. When combined with divalent europium blue phosphors, and trivalent europium red phosphors, terbium green phosphors provide high-efficiency white light in trichromatic lighting technology for standard illumination in indoor lighting.

China, Zambia, and the Democratic Republic of the Congo are the main producers of cobalt, an essential part commonly used in lithium-ion batteries and certain magnets and other green energy uses.

The point is, these materials (elements) are essential to green energy production, so a move away from a linear economy to a circular economy will support the development of shorter, more transparent, and diverse green energy supply chains that can be more easily monitored for ethical practices. The beauty of shorter supply chains is that they have a direct impact on reducing greenhouse gas emissions. Moreover, a circular economy approach for green energy equipment can help lift lives; uplift and empower the depressed, the disadvantaged, and ignored individuals; strengthen communities; and ensure that workers benefit from the transition to a new clean energy economy.

So, what is the fly in the ointment with the transitioning from a linear to a circular economy?

Good question. The answer is primarily driven by the annual need for critical metals—are they available? Simply, the critical need for rare-earth metals could endanger the transition to a green energy economy. Consider, for example, our need for indium, which is used in construction of solar panels, that is expected to increase at least 12 times before 2050—and silver, for which global demand is increasing and of which there is not expected to be enough to meets all needs (Ellen MacArthur Foundation, 2023). Certain compounds of green energy equipment require the same critical elements as many electronics components, exacerbating potential shortages. The beauty of a circular economy is that it enables countries to reliably source reused or recycled materials. Ongoing studies are attempting to study economies of scale to find ways to benefit waste handlers and recyclers to find methods to handle both e-waste and green energy waste where it makes good judgment and common sense.

The bottom line: When you get right down to it, we can say without equivocation that in a circular economy, economic activity builds and rebuilds overall system health in a truly circular fashion (ILO, 2018).

IT'S BEGINNING TO SINK IN

Shifting from a linear economy to a circular economy is not an overnight accomplishment. Some out there in la-la land may perceive a shared economy and its connection to a shared economy is a utopian dream. While it is true that a shared economy might consist of some utopian attributes where consumption involves sharing rent, borrowing goods and services from providers, it is also true that it is not imaginary. The shift will take time, money, patience, trial and error (lots of it), and most importantly, awareness.

Awareness?

Yes. Awareness that linear approaches are ineffective for solving complex problems is beginning to sink in. The movement by the private sector toward embracing the principles of circular economies is growing. The circular economy is linked to the sharing economy, which aims, from the design stage, to make better use of natural resources and preserve the environment. But be advised that a circular economy and a sharing economy are not synonymous. Let's just say the two are related—well, somewhat.

So, what is a sharing economy? First, note that the circular economy and the sharing economy overlap marginally. Also, we can say that the shared economy provides new consumer practices, motivations, and barriers that motivate consumers who convert to the shared economy and the interface of an overlapping circular economy.

OK, let's do some explaining. The sharing economy consists of three pillars: the access economy, the platform economy, and the community-based economy. An *access economy* is where various assets or services are about access over ownership; it's all about renting versus buying. The sharing economy facilitates the sharing of underused assets that are free and others which require payment of fees. The *platform economy* is all about tech-centric platforms; that is, it refers to an automation approach that relies on a single detailed automation platform to conduct all a business' automation needs. The *community-based economy* is a system that encourages local substitution—the focus is on becoming more socially, economically, culturally, and ecologically sustainable. It is the mantra of those practicing voluntary, intentional simplicity, including modern eco-village communities, Amish, and Mennonite.

Whenever I lectured to my upper college-level students on the sharing economy, a certain amount of confusion was apparent related to the various new economy models in recent headlines. To resolve and avoid this confusion, I present the following summary list of the models:

- Sharing economy—focus is on the sharing of underutilized assets, monetized or not, in ways that improve efficiency, sustainability, and community.
- Collaborative economy—focus is on collaboration of consumption, production, finance, and learning.
- On-demand economy—as its name implies, the focus is on immediate and access-based (i.e., "on-demand") provision of goods and services.
- Gig economy—focus is on independent laborers who are paid for individual tasks or appearances (aka gigs).
- Freelance economy—focus is on the workforce participant and income generation by freelancers, also known as independent workers and as self-employed.
- Access economy—focus is on access over ownership.
- Digital economy—focus is on anything powered by digital technologies.
- Peer economy—focus on what is called P2P (peer-to-peer) networks in the creation of products, funding, delivery of services, and so forth.
- Crowd economy—is a dynamic ecosystem of productive humans (i.e., powered by the crowd) who participate through a platform with a purpose to achieve mutually beneficial goals.
- Freelance economy—focus is on workforce participation and income generation by freelancers (i.e., self-employed and independent workers).

CURRENT GREEN ENERGY PRACTICES

Green energy projects increasingly combine technologies such as wind, solar, and storage (e.g., lithium-ion batteries) to offset the technology-specific occurrences at irregular intervals of variable green energy and allow for more efficient utilization and transmission capacity (USAID, 2020).

SOLAR PV: REUSE AND REFURBISH

The first generation of solar panels will reach end-of-life soon. Reuse and refurbishment will need to keep pace with the stacks of no longer effective solar panels. Although reuse and refurbishment of solar panels is uncommon, there are some emerging local markets for used solar panels. When solar panels reach their end-of-use in grid-connected setups, either they stay where they are or are replaced. Note that undamaged panels lose their efficiency over time (about 0.5% each year)—also note that no one really knows the exact life span of a solar panel because there is considerable uncertainty. At the present time, there are a few estimates that actual panel life is roughly 12.5 years—the life span could be double this number but due to breakage (mostly), the panel's life is limited. Note that PV module defects increased from 19% in 2013 to 48% in 2016, resulting in increased replacement costs (Dupont, n.d.). Defects mean early retirement of the panels. Jordan (2017) observed that module degradation depends on the type of modules: discoloration is more common in C-Si modules, whereas glass breakage is more common in thin-film modules. It is important to point out that panels placed in hot and humid climates have -average rates of degradation.

Solar panel recycling is slowly developing. One recycling plant in Yuma, Arizona, is becoming a popular shipping destination for worn-out solar panels that on any given day consist of hundreds of solar panels stacked in neat piles, waiting for their next life. At the present time, most worn-out and damaged panels are still dumped in landfills. Recognition that this needs to change is growing daily. The panels come from locations across the country. Because solar panels are built to withstand several years of harsh weather, it's difficult to break the resilient bonding that keeps them together. Using robotic suction arms assisted by workers, separate the glass without it shattering. The panels contain highly valuable materials such as copper, silver, aluminum glass, and crystalline silicon. The major hurdle potential solar panel recyclers is the struggle to pay about $30 per panel versus about $1 to a landfill. This cost to recycle or repurpose solar panels may become more attractive to recyclers if more recycling centers are constructed and prices become more competitive with landfilling (O'Malley, 2023).

ENERGY STORAGE LI-ION BATTERIES: REUSE AND REFURBISH

While there are several storage battery types, such as lead acid, sodium chemistries, flow-vanadium, flow-zinc, and other batteries, it is the lithium-ion battery (LIB) that is our focus here. Once a LIB battery is used, it can be recharged (i.e., if it can still hold a charge), be recycled, or be shifted to a different storage project—their used in two- and three-wheelers, golf carts, and buses are examples where they might be shifted to because these have lower performance demand than passenger vehicles. Note that stationary storage using LIBs is new and likely to follow EV trends. Where demand on the battery is not heavy, it can service off-grid applications, including solar home systems and community-scale grids (USAID, 2023). The point is, retired EV LIB cells and modules may be refurbished/modified for reuse in other mobile battery energy storage (BES) systems (e.g., forklifts) for reuse in stationary BES applications. Another option to be used with used LIBs is to recycle them—to recover materials from the used LIBs to manufacture new batteries or be sold into commodity markets. They also can be stored or disposed of. At the present time, life expectancy estimates for large set-up stationary LIBs for battery energy storage (BES) range from 7 to 15 years. Moreover, total installed large-scale stationary BES is expected to increase almost tenfold from 2021 to 2025 (Curtis et al., 2021; Wood Mackenzie-ESA, 2021). Passenger EVs are expected to reach 16 million units on US roads by 2030 and 46 million by 2040, and the volume of LIBs that have reached the end of their utility could reach 2 million units annually by 2040 in the United States (BloombergNEF, 2020).

Presently, 70% of cell capacity and approximately 87% of battery packs for light duty EVs are US made. Tesla and Panasonic lead battery manufacture for stationary batteries. One of the major issues or problems with the manufacture of LIBs is that the United States is heavily reliant on raw material imports from domestic manufacturing. The US Department of Interior determined that cobalt, graphite, lithium, and magnesium are critical materials essential to economic and national security. Complicating the LIB material supply is the demand for graphite, lithium, and cobalt that may increase by nearly 500% by 2050. The US Geological Survey and the US Department of Commerce found that cobalt is among the materials at the highest risk of supply chain disruption. The present movement is to replace cobalt with nickel content in battery cathode design—the problem is that within 3–7 years, if current trend persists, this will cause a shortage of nickel (Curtis et al., 2021; Wood MacKenzie-ESA, 2021).

Regarding LIB management trends, and according to anecdotal evidence, there are low volumes of retired LIBs used in mobile and stationary BES in the United States; however, first-generation EV batteries are starting to reach end-of-life and the future of the large-format LIB is becoming more uncertain and worrisome. To date, there are only a very few US-led pilot projects studying reuse of large-format LIBs—basically nothing on the commercial scale. Note that the accessibility and cost of large-format LIB recycling is regularly overshadowed by cheaper and more accessible storage and disposal options. The bottom line: Experience and various studies suggest that less than 5% of LIBs from EVs in the United States are sent to recyclers (Curtis et al., 2021; Kumagai, 2021; Warren, 2020; Gerstin, 2020; Steward et al., 2019; Collins, 2019; Kelleher Environmental, 2019; NREL, 2019; Salim et. al., 2019; DTSC, 2019; CPUC, 2019; Jacoby, 2019).

Note that for LIBs there are drivers versus barriers of a circular economy. The drivers include cost savings and increased profits; enhanced competitiveness and expanded market and employment opportunities; reduced negative environmental impacts; and reduced resource constraints.

OK, in nutshell, fashioning an explanation of what these drivers really are—what they accomplish—is called for. What the drivers do is decrease manufacturing costs and achieve additional revenue streams and tax benefits. Moreover, drivers increase consumer trust because a business' green or environmentally responsible image is enhanced. Drivers also provide opportunities for new and expanded markets, and job creation. Based on environmental aspects and concerns, these drivers result in reduced waste, reduced generation of greenhouse gases, reduction of other environmental pollutants, and reduction in total energy required to mine, transport, refine, and manufacture products. Last, drivers of a circular economy for LIBs is the conservation of high-value materials, prevention of resource constraints, and reduced import demand of raw materials (Curtis et al., 2021).

On the other side of the coin, so to speak, and according to Curtis et al., (2021), there are the barriers to a circular economy for LIBs. These barriers include current technology, infrastructure, and processes that are not optimized for efficient cost-effective refurbishment for reuse or recycling of LIBs. There is also a critical lack of information and data regarding the value of, and markets for, used LIBs and recovered LIB materials; the volume and composition of retired LIBs; the condition and characteristics of retired quality, performance, reliability, safety, and technical viability of repurposed LIBs; refurbishment and recycling technology, services, processes; costs; and infrastructure. Another issue at the present time are the unclear, complex, and varied laws and regulations applicable to reuse and recycling of LIBs that are unclear, complex, vary by jurisdiction, and often require compliance with stringent handling, storage, transport, treatment, recycling, and disposal requirements that are subject to civil and criminal liability for noncompliance. There is also a lack of economic motivation or incentive to enable collection transport, reuse, or recycling of LIBs, or to enable the design for durability, reuse, and recycling. Finally, there is low market confidence in refurbished and reused LIBs—there is inadequate consumer confidence in repurposed LIBs to support secondary markets.

THINK ABOUT IT!

Extending battery life is a way to avoid breakdown, reduce waste, and postpone recycling. Recycling LIBs used in EVs is a vital step toward making EVs sustainable because we can recover and reuse minerals already mined.

WIND ENERGY: REUSE AND REFURBISHMENT

Referring back to Figure 3.4, wind turbine equipment consists of the rotor—blades, generator, control electronics, and a gearbox—and the surrounding and supporting structure, including the steel tower, as shown in Figure 3.4. Note that some turbine designs, mostly those containing gearboxes, employ permanent magnets that typically contain critical metals primarily produced by China and Russia—they are the main suppliers of rare-earth elements at the present time.

Rare-earth magnets are strong permanent magnets, the strongest type of permanent magnets, made from alloys of rare-earth elements—on average, a permanent magnet contains almost 29% neodymium, more than 4% dysprosium, 1% boron, 66% iron, and weighs up to 4 tons. Although they have only been utilized since the 1970s, they have moved forward to a pinnacle place for practical usage in industry and elsewhere. The magnetic field typically produced by rare-earth magnets can exceed 1.4 teslas, whereas other common types of permanent magnets typically exhibit fields of 0.5 to 1 tesla—tesla is the unit of magnetic flux desist in the SI units (International System of Units).

Note that once durable steel and other alloys are magnetized, however, they retain a large part of their magnetic strength and are called *permanent magnets*. Conversely, materials that are easy to magnetize—such as soft iron and annealed silicon steel—are said to have a *high permeability*. Such materials retain only a small part of their magnetism after the magnetizing force is removed and are called *temporary magnets*.

Going large is the present trend in wind turbine construction, and because newer machines are so much larger and designed differently than they were 25 years ago, there is no reuse and refurbishment, that is, with the possible exception parts of the turbines and ancillary electric components. Instead, the industry is actively working on repowering and decommissioning with associated recycling. BNEF (2020) predicts that 15 GW of turbines will be repowered, and 19 GW will need to be removed or replaced this decade. The bottom line: Repowering typically involves replacement with new equipment.

DID YOU KNOW?

It has been estimated that there are about 14,000 wind blades that will be decommissioned by 2023; equivalent to between 40,000 and 60,000 (WindEurope, 2020).

All things wear out with time. So, what this means to the green energy planner is that there is an absolute need for a well-planned decommissioning. The truth is, decommissioning a large wind turbine is no walk in the park, no trip to the Galapagos, and not a piece of cake, so to speak. Also, the decommissioning requires multiple stakeholders. One thing that helps with the decommissioning is the guidance provided by WindEurope's guidance document *Decommissioning of Onshore Wind Turbines*. *Note*: This is an excellent guide that I have studied for some time.

OK, back to decommissioning wind turbines. In doing so, the first step in components removal are the rotor blades and hub, followed in order by the nacelle, tower, foundation, and crane assembly and access routes. To accomplish these decommissioning and dissembling steps, viewed and implemented in a circular way in which permitting formalities, project management, data requirements, risk assessment, monitoring, the restoration, a communications plan, and an environment, health, and safety plan are included, is the bottom line in decommissioning wind turbines (USAID, 2023).

NOTE

1 From F. Spellman (2022). *The Science of Wind Power* (in production). Boca Raton, FL: CRC Press.

REFERENCES

Accentur. 2015. *Waste to Wealth: Creating Advantage in Circular Economy*. Accessed 8/3/23 @ _acnmedia/ www.accenture.com/Accenture/conversion-assets/dotcom/document/glo bal/pdf/strategy_7/Accenture-wate-wealth-infographic.pdf

Bemreuter, Research. n.d. *Solar Industry Value Chain*. Accessed 8/17/2023 @ www.bemreuter. com/solar-industry/calue-chain/

BloombergNEF. 2020a. *Electric Vehicle Outlook*. Uniondale, NY.

BloombergNEF. 2020b. The afterlife of solar panels. *Bloomberg New Energy Finance*. Accessed 8/3/23 @ www.bnef.com/insights/24259

California Department of Toxic Substances Control (DTSC). 2019. *Defining Hazardous Waste*. https://dtsc.ca.gov/wp-content/uploads/sites/31/2018/05/DefiningHazardousWaste.pdf

California Public Utilities Commission (CPUC) and California Department of Resources Recycling and Recovery (Calrecycle). 2019. *Memorandum of Understanding between California Public Utilities Commission and the California Department of Resources Recycling and Recovery*. Accessed 8/5/23 @ https://bit.ly/3xq//TVU

Cascio, J. 2023. *Neodymium Is One of 17 "Rare Earth Metals."* Accessed 8/15/23 @ www.brai nyquote.com/quotes/Jamais-Cascio-560444

Circularity Gap Reporting Initiative. 2020. *Circularity Gap Report* (2020). Accessed 8/3/23 @ https://assets.web.files.com

Collins, B. 2019. *Electric Car Batteries May Enjoy Life after Death*. BloombergNEF. Accessed 8/5/23 @ www.bnef.com/core/insights/19919

Curtis, T.L., Smith, L., Buchanan, H., and Heath, G. 2021. *A Circular Economy for Lithium-Ion Batteries Used in Mobile and Stationary Energy Storage: Drivers, Barriers, Enablers, and US Policy Considerations*. Golden, CO: National Renewable Energy Laboratory. NREL/TP-6A20-77035.

Dierkes, S.J. 2003. *Dysprosium. Material Safety Data Sheet*. New York: ESPI.

Ellen MacArthur Foundation. 2019. *Completing the Picture: How the Circular Economy Tackles Climate Change*. Accessed 8/3/23 @ https://emg.thirdlight.com/link/2j2gt yton7ia

Ellen MacArthur Foundation. 2023. *What Is a Circular Economy?* Accessed 8/4/23 @ https:// ellenmacarthurfoundation.org/topics/circular-economy-introduction/overview

Energy Efficiency and Renewable Energy (EERE). 2023. *The Circular Economy*. Washington, DC: Energy Efficiency & Renewable Energy.

Gerstin, J. 2020. *The Case for Designing a Circular Battery*. Greenbiz. Accessed 8/5/23 @ www.greenbiz.com/artilcle/case-designing-circular-battery

Gorman, S. 2009. *Neodymium. Hybrid Cars Gobble-Up Rare Metals*. Reuters.

International Labor Organization (ILO). 2018. *24 Million Jobs to Open up in the Green Economy*. Accessed 8/4/23 @ www.ilo.org/global/about-the-ilo/newsroom/news/ WCMS_628644/lang--en/index

Jacoby, M. 2019. It's time to get serious about recycling lithium-ion batteries. *Chemical and Engineering News* **97**(28): 29–32.

Jordan, D.C. 2017. *Photovoltaic Failure and Degradation Modes*. Accessed 8/4/23 @ https:// dou.org/10.1002/pip.2866

Kamber, I., Bergman, A., Eich, A., Lusan, D., Steinbrecher, M., Hauptmann, N. Nordstrom, L., Katsnelson, M.L., Wagner, D., Eriksson, O.L., and Khajetoorians, A.A. 2020. *Self-Induced Spin Glass State in Elemental and Crystalline Neodymium*. Accessed 10/27/ 2021 @ https://science.sciencemag.org/content/368/6494/eaay6757

Kelleher Environmental. 2019. *Research Study on Reused and Recycling of Batteries Employees in Electric Vehicles: The Technical, Environmental, Economic, Energy and*

Cost Implications of Reusing and Recycling EV Batteries. Prepared for API by Kelleher Environmental. Accessed 8/7/23 @ https://tinyurl.com/2fxy32tn

Kumagai, J. 2021. *Lithium-Ion Battery Recycling Finally Takes off in North America and Europe*. IEEE Spectrum. Accessed 8/7/23 @ https://tinyurl.com/yn8f7m4s

Lide, D.R. 2007. Praseodymium. CRC Handbook of Chemistry and Physics. New York: CRC Press.

McGill University. 2021. Magnetic Storage. Montreal, Canada: McGill U.

National Renewable Energy Laboratory (NREL). 2019. *Competition Spurs Transformative Lithium-Ion Battery Recycling Solutions*. Accessed 8/7/23 @ www.nrel.gov/news/prog ram/2019/competition-incentivisesamerican-entrpreneurs-devlop-transformative-soluti ons-lithium-ion-battery-recycling-suppply chain.html

O'Malley, I. 2023. Hot new industry emerging. *The Virginian-Pilot*, Aug. 4.

Radboud University. 2020. *New "Whirling State of Matter Discovered: Self-Induced Spin Glass."* Accessed 10/27/2021 @ https://scietchdaily.com/new-whirling-state-of-matter-discovered-self-induced=spin-glass/

Salim, H.K., Stewart, R.A., Sahin, O., and Dudley, M. 2019. Divers barriers and enablers to end-of-life management solar photovoltaic and battery energy storage systems: A systematic literature review. *Journal of Cleaner Production* **211**: 537–554.

Sheffield Hallam University, Helena Kennedy Centre. n.d.. *In Broad Daylight: Uyghur Forced Labor and Global Solare Supply Chains*. Accessed 8/4/23 @ www.shu.ac_uk-helena-kennedy-centre-international-justice/research-and-projects//in.broad-daylight

Steward, D., Mayyas, A., and Mann, M. 2019. Economics and challenges of li-ion battery recycling from end-of-life vehicles. *Procedia Manufacturing* **33**: 272–270.

USAID. 2020. *Designing Solutions in System-Friendly Renewable Energy Competitive Procurement*. Accessed 8/4/23 @ www.usaid.gov/sites/defauly/files/documents/1865/ USAID_SURE_designing-solutions-system-friendly-renewable-energy-competitive-procurement.pdf

USAID. 2023. *Clean Energy and the Circular Economy: Opportunities for Increasing the Sustainability of Renewable Energy Value Chains*. Accessed 7/29/23 @ usaid.gov/ energy/sure

Warren, T. 2020. Running on EV: The race to solve lithium-ion battery recycling before it's too late. Fortune. Accessed 8/7/23 @ https://fortune.com/2020/01/28/lithium-ionbatt ery-recycling-electric-vehicles/

WindEurope. 2019. *Market Outlook for 2023*. Accessed 8/3/23 @ https://windeurope.org/ about-wind/reports/wind-energy-in-europe-outlook-to-2023/#download

WindEurope. 2020. *Accelerating Wind Turbine Blade Circuitry*. Accessed 8/4/23 @ https:// windeurope.org/wp-content/uploads/files/about-windeurope-accelerating-wind-turb ine-blade-circularity.pdf

Wood Mackenzie-Energy Storage Association (ESA). 2021. *US Storage Monitor: 2020 Year In Review (Full Report)*. Edinburgh, UK.

World Economic Forum (WEF). 2014. *Towards the Circular Economy: Accelerating the Scale-Up across Global Supply Chains*. Accessed 7/31/23 @www.weformum.org/agenda/ 2022/06/what-is-a-cicular-economy

15 Conclusion

Completely converting how we produce, transport, consume energy, and incorporate green energy into everyday operations is imperative to reduce global carbon dioxide emissions to net zero emissions by the middle of the twenty-first century. Note that *net zero* refers to the balance between the amount of greenhouse gases produced and the amount removed from the atmosphere. We will reach net zero when the amount added is no more than the amount taken away. Boosting and escalating implementation of green energy technological generation and integration is critical. Keep in mind that for green energy to be a truly clean energy source, solar, wind, and battery storage equipment must be manufactured, deployed, and decommissioned in a responsible, safe, and sustainable manner. A circular economy for green energy equipment can fashion a lower-emission supply chain for materials, reduce waste, create jobs, and ensure that communities and workers optimally benefit from the conversion to a clean green energy economy.

Circular economy approaches are necessary to achieve these pathways quickly, economically, and sustainably while powering economies and creating economic opportunities for more people. In a circular economy, parts and materials have multiple life cycles and reentry points into the market as they are systematically recovered, repaired, reused, refurbished, and remade into similar or other products.

OK, what this amounts to is a paradigm shift that changes how the industry approaches design to more easily and cost effectively facilitate earlier cycles before end of use and to make decommissioning and recovery more economical. And unlike the current "linear economy," this reduces the need for resources as well as supply chain risks.

The paradigm shift from conventional oil and gas to green energy sources is not, and will not be a smooth transitional shift. The most telling and pointed example for me of the rough road ahead to have a gigantic shift from hydrocarbon energy sources to green energy sources came out in one of my upper college-level environmental engineering/health courses I taught. After discussing the paradigm shift, one of my students spoke up and said, "My family bought an EV and got rid of it after a few months—my father stated a fact that when you run out of gas, a 5-gallon gas can and a nearby service station fixes that problem … but when the EV battery dies on the open road what are you to do? There is no 5-gallon can of electricity to recharge the battery."

DOI: 10.1201/9781003439059-18

Index

Note: Page numbers in *italic* refers to Figures.

A

abyssal plain, 181, *181*
access economy, 284
acetogenesis, 142
acid esterification, 128, 129
acidogenesis, 142
advective winds, 45
agricultural residues, 115
air conditioning, 219
air currents, 43, 44
air density
    correction factors, 55
    equation for, 55
algae
    as autotrophic or heterotrophic, 135, 136
    classification of, 133, *134*, 135–136
    definition of, 131–132
    farm, *138*, 138–140, *139*
    as fuel, 129–131, 138
    motility of, 133
    reproduction of, 135
    species being studied as fuel sources, 130–131
    structure of, 133, 135
    terminology of, 132–133
alkali content, 151–152, *152*
American Recovery and Reinvestment Act, 205
ammonium, 149
anaerobic digestion, *141*, 141–142, 152–153
    of agricultural wastes, 149
    dangers of, 145
    digestor operating conditions for, *147*
    equations for, 146–149
    of sewage biosolids, 143–145, *144*, *145*
    stages of, 142, *142*
anemometer, 65
angiosperms, 107
asbestos, 240
asphalt, 225
atmospheric pressure, 44–46
attenuator, 190, *190*
attic air leaks, sealing of, 238–241
attrition, 183
auxins, 109–110

B

bagasse, 156
ballast efficacy factor, 34

ballast factor, 34
ballasts, 34–35
barrier islands, 184
basement air leaks, sealing of, 242–243
batteries, 27, 62
    lithium-ion, 264
beaches, 184
BEF, *see* ballast efficacy factor
Betz, Albert, 57
Betz' law, 57, 59, *60*
    equations for, 57–59
BF, *see* ballast factor
BIG/CC, *see* biomass integrated gasification/
    combined cycle
binary-cycle power plants, 167, 168, *171*
biodiesel, 97, 113, 115
    algae-based, 129–131, 138
    characteristics of, 127, *127*
    production processes using oils and fats,
        128–129, *130*
    pros and cons of, 140
    See also biofuels
biodiesel refining, 128
bioenergy, 97
    history of, 97–98
    technological advances in, 98
bioethanol, 97
    See also biofuels
biofuels, *10*, 12, 97
    benefits of, 155–156
    vehicle consumption of, 98
Biofuels Program, 131
biogas, 141–143
    from animal waste, 149
    from landfills, 149–150, *150*
    typical contents of, *143*
biomass, *10*, 12, 97–98, 155–156
    for bioproducts, 153, *153*, *154*
    components of algal, 137
    composition of, 102–103
    definition of, 100
    methods of converting to biopower, 151
    transportation costs in recovering, 116
biomass integrated gasification/combined cycle,
    151
biopower, 151
bioproducts, 153, *153*, *154*
    classes of, 154

Printed in the United States
by Baker & Taylor Publisher Services